PRAISE FOR *EMOTIONAL DESIGN*

"Norman's analysis of people's emotional reactions to material objects is a delightful process. . . . His readers will take away insights galore about why shoppers say, 'I want that.'"

—*Booklist*

"Norman's analysis of the design elements in products such as automobiles, watches, and computers will pique the interest of many readers, not just those in the design or technology fields."

—*Publishers Weekly*

"Donald Norman's relentless and exacting exploration of the universe of everyday objects has brought him to the final frontier of design: emotions. His exquisite psychological analysis provides a solid and reliable reference and a most valuable tool."

—Paola Antonelli, Curator of Architecture
and Design, Museum of Modern Art

"Don Norman does it again! He asks the important questions and gives the right answers. I wish this insightful book had been available forty years ago so that I could have done a much better job as a designer."

—Dr. Robert Blaich, former Senior Vice President
of Corporate Design, Royal Philips Electronic

"This is a valuable book. . . . It will help the design world to do great work."

—Patrick Whitney, Director, Institute of
Design, Illinois Institute of Technology

"Amazing . . . Norman does a wonderful job making these ideas come alive."

—Daniel Bobrow, Research Fellow, Palo Alto Research Center

Emotional Design

ALSO BY DONALD A. NORMAN

The Invisible Computer

Things That Make Us Smart

Turn Signals Are the Facial Expressions of Automobiles

The Design of Everyday Things

The Psychology of Everyday Things

User Centered System Design: New Perspectives
on Human-Computer Interaction
(Edited with Stephen Draper)

Learning and Memory

Perspectives on Cognitive Science (Editor)

Human Information Processing (With Peter Lindsay)

Explorations in Cognition
(With David E. Rumelhart and the LNR Research Group)

Models of Human Memory (Editor)

Memory and Attention:
An Introduction to Human Information Processing

Emotional Design

Why We Love (*or Hate*) Everyday Things

Donald A. Norman

A MEMBER OF THE PERSEUS BOOKS GROUP
NEW YORK

Published by Basic Books,
A Member of the Perseus Books Group

Paperback edition published in 2005.
Hardcover edition published in 2004.

Books published by Basic Books are available at special discounts for bulk purchases in the United States by corporations, institutions, and other organizations. For more information, please contact the Special Markets Department at the Perseus Books Group, 11 Cambridge Center, Cambridge, MA 02142, or call 617/252-5298 or 800/255-1514, or e-mail special.markets@perseusbooks.com.

Designed by Lovedog Studio

LIBRARY OF CONGRESS CATALOGING-IN-PUBLICATION DATA
Norman, Donald A.
Emotional design : why we love (or hate) everyday things / Donald A. Norman.
p. cm.
Includes bibliographical references and index.
ISBN-10: 0-465-05135-9 (hardcover)
ISBN-13: 978-0-465-05135-9 (hardcover)
1. Emotions and cognition. 2. Design--Psychological aspects. 3. Design, Industrial—Psychological aspects. I. Title.
BF531.N67 2004
155.9'11—dc21

ISBN-10: 0-465-05136-7 (paperback)
ISBN-13: 978-0-465-05136-6 (paperback)

DHSB 21 20 19 18 17 16 15 14 13

To Julie

Contents

Emotional Design

FIGURE 0.1

An impossible teapot.

(Author's collection. Photograph by Ayman Shamma.)

FIGURE 0.2

Michael Graves's "Nanna" teapot.

So charming I couldn't resist it.

(Author's collection. Photograph by Ayman Shamma.)

FIGURE 0.3a, b, and c

The Ronnefeldt "tilting" teapot. Put leaves on the internal shelf (not visible, but just above and parallel to the ridge that can be seen running around the body of the teapot), fill with hot water, and lay the teapot on its back (figure a). As the tea darkens, tilt the pot, as in figure b. Finally, when the tea is done, stand the teapot vertically as in figure c, so the water no longer touches the leaves and the brew does not become bitter.

(Author's collection. Photographs by Ayman Shamma.)

PROLOGUE

Three Teapots

> If you want a golden rule that will fit everybody, this is it: Have nothing in your houses that you do not know to be useful, or believe to be beautiful.
>
> —*William Morris*
> *"The Beauty of Life," 1880*

I HAVE A COLLECTION OF TEAPOTS. One of them is completely unusable—the handle is on the same side as the spout. It was invented by the French artist Jacques Carelman, who called it a coffeepot: a "coffeepot for masochists." Mine is a copy of the original. A picture of it appears on the cover of my book *The Design of Everyday Things.*

The second item in my collection is the teapot called Nanna, whose unique squat and chubby nature is surprisingly appealing. The third is a complicated but practical "tilting" pot made by the German firm Ronnefeldt.

The Carelman pot is, by intent, impossible to use. The Nanna teapot, designed by the well-known architect and product designer

Michael Graves, looks clumsy but actually works rather well. The tilting pot, which I discovered while enjoying high tea at the Four Seasons Hotel in Chicago, was designed with the different stages of tea brewing in mind. To use it, I place the tea leaves on a shelf (out of sight in the pot's interior) and lay the pot on its back while the leaves steep. As the brew approaches the desired strength, I prop the pot up at an angle, partially uncovering the tea leaves. When the tea is ready, I set the pot upright, so that the leaves are no longer in contact with the tea.

Which one of these teapots do I usually use? None of the above.

I drink tea every morning. At an early hour, efficiency comes first. So, upon awakening, I pad into my kitchen and push the button on a Japanese hot pot to boil water while I spoon cut tea leaves into a little metal brewing ball. I drop the ball into my cup, fill it with boiling water, wait a few minutes for it to steep, and my tea is ready to drink. Fast, efficient, easy to clean.

Why am I so attached to my teapots? Why do I keep them out on display, in the alcove formed by the kitchen window? Even when they are not in use, they are there, visible.

I value my teapots not only for their function for brewing tea, but because they are sculptural artwork. I love standing in front of the window, comparing the contrasting shapes, watching the play of light on the varied surfaces. When I'm entertaining guests or have time to spare, I brew my tea in the Nanna teapot for its charm or in the tilting pot for its cleverness. Design is important to me, but which design I choose depends on the occasion, the context, and above all, my mood. These objects are more than utilitarian. As art, they lighten up my day. Perhaps more important, each conveys a personal meaning: each has its own story. One reflects my past, my crusade against unusable objects. One reflects my future, my campaign for beauty. And the third represents a fascinating mixture of the functional and the charming.

The story of the teapots illustrates several components of product design: usability (or lack thereof), aesthetics, and practicality. In cre-

FIGURE 0.4

Three teapots: works of art in the window above the kitchen sink.

(Author's collection. Photograph by Ayman Shamma.)

ating a product, a designer has many factors to consider: the choice of material, the manufacturing method, the way the product is marketed, cost and practicality, and how easy the product is to use, to understand. But what many people don't realize is that there is also a strong emotional component to how products are designed and put to use. In this book, I argue that the emotional side of design may be more critical to a product's success than its practical elements.

The teapots also illustrate three different aspects of design: visceral, behavioral, and reflective. *Visceral* design concerns itself with appearances. Here is where the Nanna teapot excels—I so enjoy its appearance, especially when filled with the amber hues of tea, lit from beneath by the flame of its warming candle. *Behavioral* design has to do with the pleasure and effectiveness of use. Here both the tilting teapot and my little metal ball are winners. Finally, *reflective* design considers the rationalization and intellectualization of a product. Can I tell a story about it? Does it appeal to my self-image, to my pride? I

FIGURE 0.5

The MINI Cooper S.

"It is fair to say that almost no new vehicle in recent memory has provoked more smiles." *(Courtesy of BMW AG.)*

love to show people how the tilting teapot works, explaining how the position of the pot signals the state of the tea. And, of course, the "teapot for masochists" is entirely reflective. It isn't particularly beautiful, and it's certainly not useful, but what a wonderful story it tells!

Beyond the design of an object, there is a personal component as well, one that no designer or manufacturer can provide. The objects in our lives are more than mere material possessions. We take pride in them, not necessarily because we are showing off our wealth or status, but because of the meanings they bring to our lives. A person's most beloved objects may well be inexpensive trinkets, frayed furniture, or photographs and books, often tattered, dirty, or faded. A favorite object is a symbol, setting up a positive frame of mind, a reminder of pleasant memories, or sometimes an expression of one's self. And this object always has a story, a remembrance, and something that ties us personally to this particular object, this particular thing.

Visceral, behavioral, and reflective: These three very different dimensions are interwoven through any design. It is not possible to have design without all three. But more important, note how these three components interweave both emotions and cognition.

This is so despite the common tendency to pit cognition against emotion. Whereas emotion is said to be hot, animalistic, and irrational, cognition is cool, human, and logical. This contrast comes from a long intellectual tradition that prides itself on rational, logical reasoning. Emotions are out of place in a polite, sophisticated society. They are remnants of our animal origins, but we humans must learn to rise above them. At least, that is the perceived wisdom.

Nonsense! Emotions are inseparable from and a necessary part of cognition. Everything we do, everything we think is tinged with emotion, much of it subconscious. In turn, our emotions change the way we think, and serve as constant guides to appropriate behavior, steering us away from the bad, guiding us toward the good.

Some objects evoke strong, positive emotions such as love, attachment, and happiness. In reviewing BMW's MINI Cooper car [figure 0.5], the *New York Times* observed: "Whatever one may think of the MINI Cooper's dynamic attributes, which range from very good to marginal, it is fair to say that almost no new vehicle in recent memory has provoked more smiles." The car is so much fun to look at and drive that the reviewer suggests you overlook its faults.

Several years ago, I was taking part in a radio show along with designer Michael Graves. I had just criticized one of Graves's creations, the "Rooster" teapot, as being pretty to look at, but difficult to use—to pour the water was to risk a scalding—when a listener called in who owned the Rooster. "I love my teapot," he said. "When I wake up in the morning and stumble across the kitchen to make my cup of tea, it always makes me smile." His message seemed to be: "So what if it's a little difficult to use? Just be careful. It's so pretty it makes me smile, and first thing in the morning, that's most important."

One side effect of today's technologically advanced world is that it is not uncommon to hate the things we interact with. Consider the rage and frustration many people feel when they use computers. In an article on "computer rage," a London newspaper put it this way: "It starts out with slight annoyance, then the hairs on your neck start to prickle and your hands begin to sweat. Soon you are banging your computer or yelling at the screen, and you might well end up belting the person sitting next to you."

In the 1980s, in writing *The Design of Everyday Things*, I didn't take emotions into account. I addressed utility and usability, function and form, all in a logical, dispassionate way—even though I am infuriated by poorly designed objects. But now I've changed. Why? In part because of new scientific advances in our understanding of the brain and of how emotion and cognition are thoroughly intertwined. We scientists now understand how important emotion is to everyday life, how valuable. Sure, utility and usability are important, but without fun and pleasure, joy and excitement, and yes, anxiety and anger, fear and rage, our lives would be incomplete.

Along with emotions, there is one other point as well: aesthetics, attractiveness, and beauty. When I wrote *The Design of Everyday Things*, my intention was not to denigrate aesthetics or emotion. I simply wanted to elevate usability to its proper place in the design world, alongside beauty and function. I thought that the topic of aesthetics was well-covered elsewhere, so I neglected it. The result has been the well-deserved criticism from designers: "If we were to follow Norman's prescription, our designs would all be usable—but they would also be ugly."

Usable but ugly. That's a pretty harsh judgment. Alas, the critique is valid. Usable designs are not necessarily enjoyable to use. And, as my three-teapot story indicates, an attractive design is not necessarily the most efficient. But must these attributes be in conflict? Can beauty and brains, pleasure and usability, go hand in hand?

All these questions propelled me into action. I was intrigued by the difference between my scientific self and my personal life. In science, I ignored aesthetics and emotion and concentrated on cognition. Indeed, I was one of the early workers in the fields that today are known as cognitive psychology and cognitive science. The field of usability design takes root in cognitive science—a combination of cognitive psychology, computer science, and engineering, analytical fields whose members pride themselves on scientific rigor and logical thought.

In my personal life, however, I visited art galleries, listened to and played music, and was proud of the architect-designed home in which I lived. As long as these two sides of my life were separate, there wasn't any conflict. But early in my career, I experienced a surprising challenge from an unlikely source: the use of color monitors for computers.

In the early years of the personal computer, color displays were unheard of. Most of the display screens were black and white. Sure, the very first Apple Computer, the Apple II, could display color, but for games: any serious work done on the Apple II was done in black and white, usually white text on a black background. In the early 1980s, when color screens were first introduced to the world of personal computers, I had trouble understanding their appeal. In those days, color was primarily used to highlight text or to add superfluous decoration to the screen. From a cognitive point of view, color added no value that shading could not provide. But businesses insisted on buying color monitors at added cost, despite their having no scientific justification. Obviously, color was fulfilling some need, but one we could not measure.

I borrowed a color monitor to see what all the fuss was about. I was soon convinced that my original assessment had been correct: color added no discernible value for everyday work. Yet I refused to give up the color display. My reasoning told me that color was unimportant, but my emotional reaction told me otherwise.

Notice the same phenomenon in movies, television, and newspapers. At first, all movies were in black and white. So, too, was television. Movie makers and television manufacturers resisted the introduction of color because it added huge costs with little discernible gain. After all, a story is a story—what difference does color make? But would you go back to black-and-white TV or movies? Today, the only time something is filmed in black and white is for artistic, aesthetic reasons: The lack of full color makes a strong emotional statement. The same lesson has not fully transferred to newspapers and books. Everyone agrees that color is usually preferred, but whether the benefits are sufficient to overcome the additional costs it entails is hotly debated. Although color has crept into the pages of newspapers, most of the photographs and advertisements are still in black and white. So, too, with books: The photographs in this book are all in black and white, even though the originals are in color. In most books, the only place color appears is on the cover—presumably to lure you into purchasing the book—but once you have purchased it, the color is thought to have no further use.

The problem is that we still let logic make decisions for us, even though our emotions are telling us otherwise. Business has come to be ruled by logical, rational decision makers, by business models and accountants, with no room for emotion. Pity!

We cognitive scientists now understand that emotion is a necessary part of life, affecting how you feel, how you behave, and how you think. Indeed, emotion makes you smart. That's the lesson of my current research. Without emotions, your decision-making ability would be impaired. Emotion is always passing judgments, presenting you with immediate information about the world: here is potential danger, there is potential comfort; this is nice, that bad. One of the ways by which emotions work is through neurochemicals that bathe particular brain centers and modify perception, decision making, and behavior. These neurochemicals change the parameters of thought.

The surprise is that we now have evidence that aesthetically pleasing objects enable you to work better. As I shall demonstrate, products and systems that make you feel good are easier to deal with and produce more harmonious results. When you wash and polish your car, doesn't it seem to drive better? When you bathe and dress up in clean, fancy clothes, don't you feel better? And when you use a wonderful, well-balanced, aesthetically pleasing garden or woodworking tool, tennis racket or pair of skis, don't you perform better?

Before I go on, let me interject a technical comment: I am talking here about affect, not just emotion. A major theme of this book is that much of human behavior is subconscious, beneath conscious awareness. Consciousness comes late, both in evolution and also in the way the brain processes information; many judgments have already been determined before they reach consciousness. Both affect and cognition are information-processing systems, but they have different functions. The affective system makes judgments and quickly helps you determine which things in the environment are dangerous or safe, good or bad. The cognitive system interprets and makes sense of the world. Affect is the general term for the judgmental system, whether conscious or subconscious. Emotion is the conscious experience of affect, complete with attribution of its cause and identification of its object. The queasy, uneasy feeling you might experience, without knowing why, is affect. Anger at Harry, the used-car salesman, who overcharged you for an unsatisfactory vehicle, is emotion. You are angry at something—Harry—for a reason. Note that cognition and affect influence one another: some emotions and affective states are driven by cognition, while affect often impacts cognition.

Let's look at a simple example. Imagine a long and narrow plank ten meters long and one meter wide. Place it on the ground. Can you walk on it? Of course. You can jump up and down, dance, and even walk along with your eyes shut. Now prop the plank up so that it is three meters off the ground. Can you walk on it? Yes, although you proceed more carefully.

What if the plank were a hundred meters in the air? Most of us wouldn't dare go near it, even though the act of walking along it and maintaining balance should be no more difficult than when the plank is on the ground. How can a simple task suddenly become so difficult? The reflective part of your mind can rationalize that the plank is just as easy to walk on at a height as on the ground, but the automatic, lower visceral level controls your behavior. For most people, the visceral level wins: fear dominates. You may try to justify your fear by stating that the plank might break, or that, because it is windy, you might be blown off. But all this conscious rationalization comes after the fact, after the affective system has released its chemicals. The affective system works independently of conscious thought.

Finally, affect and emotion are crucial for everyday decision making. The neuroscientist Antonio Damasio studied people who were perfectly normal in every way except for brain injuries that impaired their emotional systems. As a result, despite their appearance of normality, they were unable to make decisions or function effectively in the world. While they could describe exactly how they should have been functioning, they couldn't determine where to live, what to eat, and what products to buy and use. This finding contradicts the common belief that decision making is the heart of rational, logical thought. But modern research shows that the affective system provides critical assistance to your decision making by helping you make rapid selections between good and bad, reducing the number of things to be considered.

People without emotions, as in Damasio's study, are often unable to choose between alternatives, especially if each choice appears equally valid. Do you want to come in for your appointment on Monday or Tuesday? Do you want rice or baked potato with your food? Simple choices? Yes, perhaps too simple: there is no rational way to decide. This is where affect is useful. Most of us just decide on something, but if asked why, often don't know: "I just felt like it," one might reply. A decision has to "feel good," or else it is rejected, and such feeling is an expression of emotion.

The emotional system is also tightly coupled with behavior, preparing the body to respond appropriately to a given situation. This is why you feel tense and edgy when anxious. The "queasy" or "knotted" feelings in your gut are not imaginary—they are real manifestations of the way that emotions control your muscle systems and, yes, even your digestive system. Thus, pleasant tastes and smells cause you to salivate, to inhale and ingest. Unpleasant things cause the muscles to tense as preparation for a response. A bad taste causes the mouth to pucker, food to be spit out, the stomach muscles to contract. All of these reactions are part of the experience of emotion. We literally *feel* good or bad, relaxed or tense. Emotions are judgmental, and prepare the body accordingly. Your conscious, cognitive self observes those changes. Next time you feel good or bad about something, but don't know why, listen to your body, to the wisdom of its affective system.

Just as emotions are critical to human behavior, they are equally critical for intelligent machines, especially autonomous machines of the future that will help people in their daily activities. Robots, to be successful, will have to have emotions (a topic I discuss in more detail in chapter 6). Not necessarily the same as human emotions, these will be emotions nonetheless, ones tailored to the needs and requirements of a robot. Furthermore, the machines and products of the future may be able to sense human emotions and respond accordingly. Soothe you when you are upset, humor you, console you, play with you.

As I've said, cognition interprets and understands the world around you, while emotions allow you to make quick decisions about it. Usually, you react emotionally to a situation before you assess it cognitively, since survival is more important than understanding. But sometimes cognition comes first. One of the powers of the human mind is its ability to dream, to imagine, and to plan for the future. In this creative soaring of the mind, thought and cognition unleash emotion, and are in turn changed themselves. To explain how this comes about, let me now turn to the science of affect and emotion.

PART ONE

The Meaning of Things

CHAPTER ONE

Attractive Things Work Better

NOAM TRACTINSKY, AN ISRAELI SCIENTIST, WAS puzzled. Attractive things certainly should be preferred over ugly ones, but why would they work better? Yet in the early 1990s, two Japanese researchers, Masaaki Kurosu and Kaori Kashimura, claimed just that. They studied different layouts of controls for ATMs, automated teller machines that allow us to perform simple banking tasks any time of the day or night. All versions of the ATMs were identical in function, the number of buttons, and how they operated, but some had the buttons and screens arranged attractively, the others unattractively. Surprise! The Japanese found that the attractive ones were perceived to be easier to use.

Tractinsky was suspicious. Maybe the experiment had flaws. Or perhaps the result could be true of Japanese, but certainly not of Israelis. "Clearly," said Tractinsky, "aesthetic preferences are culturally dependent." Moreover, he continued, "Japanese culture is known for its aesthetic tradition," but Israelis? Nah, Israelis are action-oriented—they don't care about beauty. So Tractinsky redid the experiment. He got the ATM layouts from Kurosu and Kashimura, translated them from Japanese into Hebrew, and designed a new experiment, with rigorous methodological controls. Not only did he replicate the Japanese findings, but—contrary to his belief that usability and aesthetics "*were not expected* to correlate"—the results were stronger in Israel than in Japan. Tractinsky was so surprised that he put that phrase "*were not expected*" in italics, an unusual thing to do in a scientific paper, but appropriate, he felt, given the unexpected conclusion.

In the early 1900s, Herbert Read, who wrote numerous books on art and aesthetics, stated, "it requires a somewhat mystical theory of aesthetics to find any necessary connection between beauty and function," and that belief is still common today. How could aesthetics affect how easy something is to use? I had just started a research project examining the interaction of affect, behavior, and cognition, but Tractinsky's results bothered me—I couldn't explain them. Still, they were intriguing, and they supported my own personal experiences, some of which I described in the prologue. As I pondered the experimental results, I realized they fit with the new framework that my research collaborators and I were constructing as well as with new findings in the study of affect and emotion. Emotions, we now know, change the way the human mind solves problems—the emotional system changes how the cognitive system operates. So, if aesthetics would change our emotional state, that would explain the mystery. Let me explain.

Until recently, emotion was an ill-explored part of human psychology. Some people thought it an evolutionary leftover from our animal origins. Most thought of emotions as a problem to be overcome by rational, logical thinking. And most of the research focused upon negative emotions such as stress, fear, anxiety, and anger. Modern work has completely reversed this view. Science now knows that evolutionarily more advanced animals are more emotional than primitive ones, the human being the most emotional of all. Moreover, emotions play a critical role in daily lives, helping assess situations as good or bad, safe or dangerous. As I discussed in the prologue, emotions aid in decision making. Positive emotions are as important as negative ones—positive emotions are critical to learning, curiosity, and creative thought, and today research is turning toward this dimension. One finding particularly intrigued me: The psychologist Alice Isen and her colleagues have shown that being happy broadens the thought processes and facilitates creative thinking. Isen discovered that when people were asked to solve difficult problems, ones that required unusual "out of the box" thinking, they did much better when they had just been given a small gift—not much of a gift, but enough to make them feel good. When you feel good, Isen discovered, you are better at brainstorming, at examining multiple alternatives. And it doesn't take much to make people feel good. All Isen had to do was ask people to watch a few minutes of a comedy film or receive a small bag of candy.

We have long known that when people are anxious they tend to narrow their thought processes, concentrating upon aspects directly relevant to a problem. This is a useful strategy in escaping from danger, but not in thinking of imaginative new approaches to a problem. Isen's results show that when people are relaxed and happy, their thought processes expand, becoming more creative, more imaginative.

These and related findings suggest the role of aesthetics in product design: attractive things make people feel good, which in turn makes them think more creatively. How does that make something easier to use? Simple, by making it easier for people to find solutions to the problems they encounter. With most products, if the first thing you try fails to produce the desired result, the most natural response is to try again, only with more effort. In today's world of computer-controlled products, doing the same operation over again is very unlikely to yield better results. The correct response is to look for alternative solutions. The tendency to repeat the same operation over again is especially likely for those who are anxious or tense. This state of negative affect leads people to focus upon the problematic details, and if this strategy fails to provide a solution, they get even more tense, more anxious, and increase their concentration upon those troublesome details. Contrast this behavior with those who are in a positive emotional state, but encountering the same problem. These people are apt to look around for alternative approaches, which is very likely to lead to a satisfying end. Afterward, the tense and anxious people will complain about the difficulties whereas the relaxed, happy ones will probably not even remember them. In other words, happy people are more effective in finding alternative solutions and, as a result, are tolerant of minor difficulties. Herbert Read thought we would need a mystical theory to connect beauty and function. Well, it took one hundred years, but today we have that theory, one based in biology, neuroscience, and psychology, not mysticism.

Human beings have evolved over millions of years to function effectively in the rich and complex environment of the world. Our perceptual systems, our limbs, the motor system—which means the control of all our muscles—everything has evolved to make us function better in the world. Affect, emotion, and cognition have also evolved to interact with and complement one another. Cognition interprets the world, leading to increased understanding and knowledge. Affect, which includes emotion, is a system of judging what's good or bad, safe or dangerous. It makes value judgments, the better to survive.

The affective system also controls the muscles of the body and, through chemical neurotransmitters, changes how the brain functions. The muscle actions get us ready to respond, but they also serve as signals to others we encounter, which provides yet another powerful role of emotion as communication: our body posture and facial expression give others clues to our emotional state. Cognition and affect, understanding and evaluation—together they form a powerful team.

• • •

Three Levels of Processing: Visceral, Behavioral, and Reflective

Human beings are, of course, the most complex of all animals, with accordingly complex brain structures. A lot of preferences are present at birth, part of the body's basic protective mechanisms. But we also have powerful brain mechanisms for accomplishing things, for creating, and for acting. We can be skilled artists, musicians, athletes, writers, or carpenters. All this requires a much more complex brain structure than is involved in automatic responses to the world. And finally, unique among animals, we have language and art, humor and music. We are conscious of our role in the world and we can reflect upon past experiences, the better to learn; toward the future, the better to be prepared; and inwardly, the better to deal with current activities.

My studies of emotion, conducted with my colleagues Andrew Ortony and William Revelle, professors in the Psychology Department at Northwestern University, suggest that these human attributes result from three different levels of the brain: the automatic, prewired layer, called the *visceral level*; the part that contains the brain processes that control everyday behavior, known as the *behavioral level*; and the contemplative part of the brain, or the *reflective level*. Each level plays a different role in the total functioning of people. And, as I discuss in detail in chapter 3, each level requires a different style of design.

The three levels in part reflect the biological origins of the brain, starting with primitive one-celled organisms and slowly evolving to more complex animals, to the vertebrates, the mammals, and finally, apes and humans. For simple animals, life is a continuing set of threats and opportunities, and an animal must learn how to respond appropriately to each. The basic brain circuits, then, are really response mechanisms: analyze a situation and respond. This system is tightly coupled to the animal's muscles. If something is bad or dangerous, the muscles tense in preparation for running, attacking, or freezing. If something

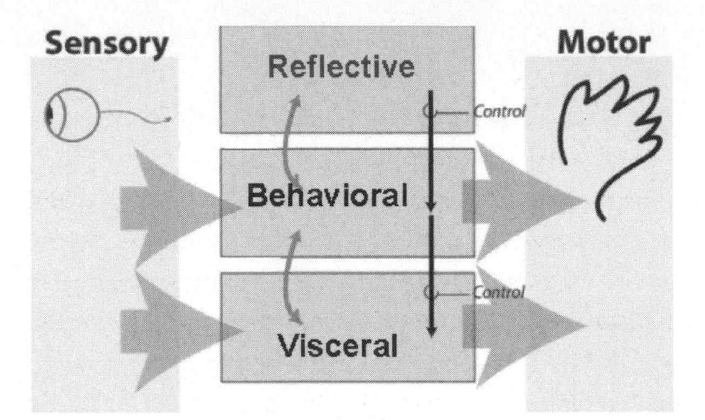

FIGURE 1.1

Three levels of processing: Visceral, Behavioral, and Reflective. The visceral level is fast: it makes rapid judgments of what is good or bad, safe or dangerous, and sends appropriate signals to the muscles (the motor system) and alerts the rest of the brain. This is the start of affective processing. These are biologically determined and can be inhibited or enhanced through control signals from above. The behavioral level is the site of most human behavior. Its actions can be enhanced or inhibited by the reflective layer and, in turn, it can enhance or inhibit the visceral layer. The highest layer is that of reflective thought. Note that it does not have direct access either to sensory input or to the control of behavior. Instead it watches over, reflects upon, and tries to bias the behavioral level.

(Modified from a figure by Daniel Russell for Norman, Ortony, & Russell, 2003.)

is good or desirable, the animal can relax and take advantage of the situation. As evolution continued, the circuits for analyzing and responding improved and became more sophisticated. Put a section of wire mesh fence between an animal and some desirable food: a chicken is likely to be stuck forever, straining at the fence, but unable to get to the food; a dog simply runs around it. Human beings have an even more developed set of brain structures. They can reflect upon their experiences and communicate them to others. Thus, not only do we walk around fences to get to our goals, but we can then think back about the experience—reflect upon it—and decide to move the fence or the food, so we don't have to walk around the next time. We can also tell other people about the problem, so they will know what to do even before they get there.

Animals such as lizards operate primarily at the visceral level. This is the level of fixed routines, where the brain analyzes the world and responds. Dogs and other mammals, however, have a higher level of analysis, the behavioral level, with a complex and powerful brain that can analyze a situation and alter behavior accordingly. The behavioral level in human beings is especially valuable for well-learned, routine operations. This is where the skilled performer excels.

At the highest evolutionary level of development, the human brain can think about its own operations. This is the home of reflection, of conscious thought, of the learning of new concepts and generalizations about the world.

The behavioral level is not conscious, which is why you can successfully drive your automobile subconsciously at the behavioral level while consciously thinking of something else at the reflective level. Skilled performers make use of this facility. Thus, skilled piano players can let their fingers play automatically while they reflect upon the higher-order structure of the music. This is why they can hold conversations while playing and why performers sometimes lose their place in the music and have to listen to themselves play to find out where they are. That is, the reflective level was lost, but the behavioral level did just fine.

Now let's look at some examples of these three levels in action: riding a roller coaster; chopping and dicing food with a sharp, balanced knife and a solid cutting board; and contemplating a serious work of literature or art. These three activities impact us in different ways. The first is the most primitive, the visceral reaction to falling, excessive speed, and heights. The second, the pleasure of using a good tool effectively, refers to the feelings accompanying skilled accomplishment, and derives from the behavioral level. This is the pleasure any expert feels when doing something well, such as driving a difficult course or playing a complex piece of music. This behavioral pleasure, in turn, is different from that provided by serious literature or art,

FIGURE 1.2

People pay money to get scared.

The roller coaster pits one level of affect—the visceral sense of fear—against another level—the reflective pride of accomplishment.

(Photograph by Bill Varie. © 2001 Corbis, all rights reserved.)

whose enjoyment derives from the reflective level, and requires study and interpretation.

Most interesting of all is when one level plays off of another, as in the roller coaster. If the roller coaster is so frightening, why is it so popular? There are at least two reasons. First, some people seem to love fear itself: they enjoy the high arousal and increased adrenaline rush that accompanies danger. The second reason comes from the feelings that follow the ride: the pride in conquering fear and of being able to brag about it to others. In both cases, the visceral angst competes with the reflective pleasure—not always successfully, for many people refuse to go on those rides or, having done it once, refuse to do it again. But this adds to the pleasure of those who do go on the ride: their self image is enhanced because they have dared do an action that others reject.

Focus and Creativity

The three levels interact with one another, each modulating the others. When activity is initiated from the lowest, visceral levels, it is called "bottom-up." When the activity comes from the highest, reflective level, it is called "top-down" behavior. These terms come from the standard way of showing the processing structures of the brain, with the bottom layers associated with interpreting sensory inputs to the body and the top layers associated with higher thought processes, much as I illustrated in Figure 1.1. Bottom-up processes are those driven by perception whereas top-down are driven by thought. The brain changes its manner of operation when bathed in the liquid chemicals called neurotransmitters. A neurotransmitter does what its name implies: It changes how neurons transmit neural impulses from one nerve cell to another (that is, across synapses). Some neurotransmitters enhance transmission, some inhibit it. See, hear, feel, or otherwise sense the environment, and the affective system passes judgment, alerting other centers in the brain, and releasing neurotransmitters appropriate to the affective state. That's bottom-up activation. Think something at the reflective level and the thoughts are transmitted to the lower levels which, in turn, triggers neurotransmitters.

The result is that everything you do has both a cognitive and an affective component—cognitive to assign meaning, affective to assign value. You cannot escape affect: it is always there. More important, the affective state, whether positive or negative affect, changes how we think.

When you are in a state of negative affect, feeling anxious or endangered, the neurotransmitters focus the brain processing. Focus refers to the ability to concentrate upon a topic, without distraction, and then to go deeper and deeper into the topic until some resolution is reached. Focus also implies concentration upon the details. It is very important for survival, which is where negative affect plays a major role. Whenever your brain detects something that might be dangerous, whether through visceral or reflective processing, your affective system acts to tense muscles in preparation for action and to alert the behavioral and reflective levels to stop and concentrate upon the problem. The neurotransmitters bias the brain to focus upon the problem and avoid distractions. This is just what you need to do in order to deal with danger.

When you are in a state of positive affect, the very opposite actions take place. Now, neurotransmitters broaden the brain processing, the muscles can relax, and the brain attends to the opportunities offered by the positive affect. The broadening means that you are now far less focused, and far more likely to be receptive to interruptions and to attending to any novel idea or event. Positive affect arouses curiosity, engages creativity, and makes the brain into an effective learning organism. With positive affect, you are more likely to see the forest than the trees, to prefer the big picture and not to concentrate upon details. On the other hand, when you are sad or anxious, feeling negative affect, you are more likely to see the trees before the forest, the details before the big picture.

What role do these states have in design? First, someone who is relaxed, happy, in a pleasant mood, is more creative, more able to overlook and cope with minor problems with a device—especially if it's fun to work with. Recall the reviewer of the Mini Cooper automobile, quoted in the prologue, who recommended that the car's faults be ignored because it was so much fun. Second, when people are anxious, they are more focused, so where this is likely to be the case, the designer must pay special attention to ensure that all the information required to do the task is continually at hand, readily visible, with clear and unambiguous feedback about the operations that the device is performing. Designers can get away with more if the product is fun and enjoyable. Things intended to be used under stressful situations require a lot more care, with much more attention to detail.

One interesting effect of the differences in thought processes of the two states is the impact upon the design process itself. Design—and for that matter, most problem solving—requires creative thinking followed by a considerable period of concentrated, focused effort. In the first case, creativity, it is good for the designer to be relaxed, in a good mood. Thus, in brainstorming sessions, it is common to warm up by telling jokes and playing games. No criticism is allowed because it would raise the level of anxiety among the participants. Good brainstorming and unusual, creative thinking require the relaxed state induced by positive affect.

Once the creative stage is completed, the ideas that have been generated have to be transformed into real products. Now the design team must exert considerable attention to detail. Here, focus is essential. One way to do this is through deadlines just slightly shorter than feel comfortable. Here is the time for the concentrated focus that negative affect produces. This is one reason people often impose deadlines on themselves, and then announce those deadlines to others so as to make them real. Their anxiety helps them get the work done.

It is tricky to design things that must accommodate both creative thinking and focus. Suppose the design task is to build a control room for operators of a plant—think of a nuclear power plant or a large chemical-processing plant, but the same lessons apply to many manufacturing and production facilities. The design is meant to enhance some critical procedure or function—say to enable control room operators to watch over a plant and solve problems as they arise—so it is probably best to have a neutral or a slightly negative affect to keep people aroused and focused. This calls for an attractive, pleasant environment so that in normal monitoring, the operators are creative and open to explore new situations. Once some plant parameter approaches a dangerous level, however, the design should change its stance, yielding a negative affect that will keep the operators focused upon the task at hand.

How do you design something so that it can change from invoking a positive affect to invoking a negative one? There are several ways. One is through the use of sound. The visual appearance of the plant can be positive and enjoyable. During normal operation, it is even possible to play light background music, unless the control room is located where the sounds of the plant operating can be used to indicate its state. But as soon as any problem exists, the music should go away and alarms should start to sound. Buzzing, ringing alarms are negative and anxiety producing, so their presence alone might do the trick. Indeed, the problem is not to overdo it: too much anxiety produces a phenomenon known as "tunnel vision," where the people become so focused they may fail to see otherwise obvious alternatives.

The dangers of too much focus are well known to people who study accidents. Thus, special design and training is required of people if we want them to perform well under high stress. Basically, because of the extreme focus and tunnel vision induced by high anxiety, the situation has to be designed to minimize the need for creative thought. That's why professionals are trained over and over again in accident scenarios, through training exercises and simulators, so that if a real incident occurs, they will have experienced it so many times in training that their responses follow automatically. But this training works only if the training is repeated frequently and performance is tested. In commercial aviation, the pilots and crew are well trained, but the passengers are not. Even though frequent fliers continually hear and see the instructions on how to escape the airplane in case of fire or crash, they sit passively, only partially attentive. They are not apt to remember them in an emergency.

"Fire," yells someone in a theater. Immediately everyone stampedes toward the exits. What do they do at the exit door? Push. If the door doesn't open, they push harder. But what if the door opens inward and must be pulled, not pushed? Highly anxious, highly focused people are very unlikely to think of pulling.

When under high anxiety—high negative affect—people focus upon escape. When they reach the door, they push. And when this fails, the natural response is to push even harder. Countless people have died as a result. Now, fire laws require what is called "panic hardware." The doors of auditoriums have to open outward, and they must open whenever pressure is applied.

Similarly, designers of exit stairways have to block any direct path from the ground floor to those floors below it. Otherwise, people using a stairway to escape a fire are likely to miss the ground floor and continue all the way into the basement—and some buildings have several levels of basements—to end up trapped.

The Prepared Brain

Although the visceral level is the simplest and most primitive part of the brain, it is sensitive to a very wide range of conditions. These are genetically determined, with the conditions evolving slowly over the time course of evolution. They all share one property, however: the condition can be recognized simply by the sensory information. The visceral level is incapable of reasoning, of comparing a situation with past history. It works by what cognitive scientists call "pattern matching." What are people genetically programmed for? Those situations and objects that, throughout evolutionary history, offer food, warmth, or protection give rise to positive affect. These conditions include:

warm, comfortably lit places,
temperate climate,
sweet tastes and smells,
bright, highly saturated hues,
"soothing" sounds and simple melodies and rhythms,
harmonious music and sounds,
caresses,
smiling faces,
rhythmic beats,
"attractive" people,
symmetrical objects,
rounded, smooth objects,
"sensuous" feelings, sounds, and shapes.

Similarly, here are some of the conditions that appear to produce automatic negative affect:

heights,
sudden, unexpected loud sounds or bright lights,
"looming" objects (objects that appear to be about to hit the observer),
extreme hot or cold,
darkness,
extremely bright lights or loud sounds,
empty, flat terrain (deserts),
crowded dense terrain (jungles or forests),
crowds of people,
rotting smells, decaying foods
bitter tastes,
sharp objects,
harsh, abrupt sounds,
grating and discordant sounds,
misshapen human bodies,
snakes and spiders,
human feces (and its smell),
other people's body fluids,
vomit.

These lists are my best guess about what might be automatically programmed into the human system. Some of the items are still under dispute; others will probably have to be added. Some are politically incorrect in that they appear to produce value judgments on dimensions society has deemed to be irrelevant. The advantage human beings have over other animals is our powerful reflective level that enables us to overcome the dictates of the visceral, pure biological level. We can overcome our biological heritage.

Note that some biological mechanisms are only predispositions rather than full-fledged systems. Thus, although we are predisposed to be afraid of snakes and spiders, the actual fear is not present in all people: it needs to be triggered through experience. Although human language comes from the behavioral and reflective levels, it provides a good example of how biological predispositions mix with experience. The human brain comes ready for language: the architecture of the brain, the way the different components are structured and interact, constrains the very nature of language. Children do not come into the world with language, but they do come predisposed and ready. That is the biological part. But the particular language you learn, and the accent with which you speak it, are determined through experience. Because the brain is prepared to learn language, everyone does so unless they have severe neurological or physical deficits. Moreover, the learning is automatic: we may have to go to school to learn to read and write, but not to listen and speak. Spoken language—or signing, for those who are deaf—is natural. Although languages differ, they all follow certain universal regularities. But once the first language has been learned, it highly influences later language acquisition. If you have ever tried to learn a second language beyond your teenage years, you know how different it is from learning the first, how much harder, how reflective and conscious it seems compared to the subconscious, relatively effortless experience of learning the first language. Accents are the hardest thing to learn for the older language-learner, so that people who learn a language later in life may be completely fluent in their speech, understanding, and writing, but maintain the accent of their first language.

Tinko and *losse* are two words in the mythical language Elvish, invented by the British philologist J. R. R. Tolkien for his trilogy, *The Lord of the Rings*. Which of the words "*tinko*" and "*losse*" means "metal," which "snow"? How could you possibly know? The surprise is that when forced to guess, most people can get the choices right, even if they have never read the books, never experienced the words. *Tinko* has two hard, "plosive" sounds—the "t" and the "k." *Losse* has soft, liquid sounds, starting with the "l" and continuing through the vowels and the sibilant "ss." Note the similar pattern in the English words where the hard "t" in "metal" contrasts with the soft sounds of "snow." Yes, in Elvish, *tinko* is metal and *losse* is snow.

The Elvish demonstration points out the relationship between the sounds of a language and the meaning of words. At first glance, this sounds nonsensical—after all, words are arbitrary. But more and more evidence piles up linking sounds to particular general meanings. For instance, vowels are warm and soft: *feminine* is the term frequently used. Harsh sounds are, well, harsh—just like the word "harsh" itself and the "sh" sound in particular. Snakes hiss and slither; and note the sibilants, the hissing of the "s" sounds. Plosives, sounds caused when the air is stopped briefly, then released—explosively—are hard, metallic; the word "masculine" is often applied to them. The "k" of "mosquito" and the "p" in "happy" are plosive. And, yes, there is evidence that word choices are not arbitrary: a sound symbolism governs the development of a language. This is another instance where artists, poets in this case, have long known the power of sounds to evoke affect and emotions within the readers of—or, more accurately, listeners to—poetry.

All these prewired mechanisms are vital to daily life and our interactions with people and things. Accordingly, they are important for design. While designers can use this knowledge of the brain to make designs more effective, there is no simple set of rules. The human mind is incredibly complex, and although all people have basically the same form of body and brain, they also have huge individual differences.

Emotions, moods, traits, and personality are all aspects of the different ways in which people's minds work, especially along the affective, emotional domain. Emotions change behavior over a relatively short term, for they are responsive to the immediate events. Emotions last for relatively short periods—minutes or hours. Moods are longer lasting, measured perhaps in hours or days. Traits are very long-lasting, years or even a lifetime. And personality is the particular collection of traits of a person that last a lifetime. But all of these are changeable as well. We all have multiple personalities, emphasizing some traits when with families, a different set when with friends. We all change our operating parameters to be appropriate for the situation we are in.

Ever watch a movie with great enjoyment, then watch it a second time and wonder what on earth you saw in it the first time? The same phenomenon occurs in almost all aspects of life, whether in interactions with people, in a sport, a book, or even a walk in the woods. This phenomenon can bedevil the designer who wants to know how to design something that will appeal to everyone: One person's acceptance is another one's rejection. Worse, what is appealing at one moment may not be at another.

The source of this complexity can be found in the three levels of processing. At the visceral level, people are pretty much the same all over the world. Yes, individuals vary, so although almost everyone is born with a fear of heights, this fear is so extreme in some people that they cannot function normally—they have acrophobia. Yet others have only mild fear, and they can overcome it sufficiently to do rock climbing, circus acts, or other jobs that have them working high in the air.

The behavioral and reflective levels, however, are very sensitive to experiences, training, and education. Cultural views have huge impact here: what one culture finds appealing, another may not. Indeed, teenage culture seems to dislike things solely because adult culture likes them.

So what is the designer to do? In part, that is the theme of the rest of the book. But the challenges should be thought of as opportunities. Designers will never lack for things to do, for new approaches to explore.

Chapter Two

The Multiple Faces of Emotion and Design

AFTER DINNER, WITH A GREAT FLOURISH, my friend Andrew brought out a lovely leather box. "Open it," he said, proudly, "and tell me what you think."

I opened the box. Inside was a gleaming stainless-steel set of old mechanical drawing instruments: dividers, compasses, extension arms for the compasses, an assortment of points, lead holders, and pens that could be fitted onto the dividers and compasses. All that was missing was the T square, the triangles, and the table. And the ink, the black India ink.

"Lovely," I said. "Those were the good old days, when we drew by hand, not by computer."

Our eyes misted as we fondled the metal pieces.

"But you know," I went on, "I hated it. My tools always slipped, the point moved before I could finish the circle, and the India ink—ugh, the India ink—it always blotted before I could finish a diagram. Ruined it! I used to curse and scream at it. I once spilled the whole bottle all over the drawing, my books, and the table. India ink doesn't wash off. I hated it. Hated it!"

"Yeah," said Andrew, laughing, "you're right. I forgot how much I hated it. Worst of all was too much ink on the nibs! But the instruments are nice, aren't they?"

"Very nice," I said, "as long as we don't have to use them."

THIS STORY shows the several levels of the cognitive and emotional system—visceral, behavioral, and reflective—at work, fighting among themselves. First, the most basic visceral level responds with pleasure to seeing the well-designed leather case and gleaming stainless-steel instruments and to feeling their comfortable heft. That visceral response is immediate and positive, triggering the reflective system to think back about the past, many decades ago, "the good old days," when my friend and I actually used those tools. But the more we reflect upon the past, the more we remember the actual negative experiences, and herein lies the conflict with the initial visceral reaction.

We recall how badly we actually performed, how the tools were never completely under control, sometimes causing us to lose hours of work. Now, in each of us, visceral is pitted against reflection. The sight of the classic tools is attractive, but the memory of their use is negative. Because the power of emotion fades with time, the negative affect generated by our memories doesn't overcome the positive affect generated by the sight of the instruments themselves.

This conflict among different levels of emotion is common in design: Real products provide a continual set of conflicts. A person interprets an experience at many levels, but what appeals at one may not at another. A successful design has to excel at all levels. While logic might imply, for example, that it is bad business to scare customers, amusement and theme parks have many customers for rides and haunted houses designed to scare. But the scaring occurs in a safe, reassuring environment.

The design requirements for each level differ widely. The visceral

FIGURE 2.1

Sky diving: An innate fear of heights or a pleasurable experience?

(Rocky Point Pictures; courtesy of Terry Schumacher.)

level is pre-consciousness, pre-thought. This is where appearance matters and first impressions are formed. Visceral design is about the initial impact of a product, about its appearance, touch, and feel.

The behavioral level is about use, about experience with a product. But experience itself has many facets: function, performance, and usability. A product's function specifies what activities it supports, what it is meant to do—if the functions are inadequate or of no interest, the product is of little value. Performance is about how well the product does those desired functions—if the performance is inadequate, the product fails. Usability describes the ease with which the user of the product can understand how it works and how to get it to perform. Confuse or frustrate the person who is using the product and negative emotions result. But if the product does what is needed, if it is fun to use and easy to satisfy goals with it, then the result is warm, positive affect.

It is only at the reflective level that consciousness and the highest levels of feeling, emotions, and cognition reside. It is only here that the full impact of both thought and emotions are experienced. At the lower visceral and behavioral levels, there is only affect, but without interpretation or consciousness. Interpretation, understanding, and reasoning come from the reflective level.

Of the three levels, the reflective one is the most vulnerable to variability through culture, experience, education, and individual differences. This level can also override the others. Hence, one person's liking for otherwise distasteful or frightening visceral experiences that might repel others, or another's intellectual dismissal of designs others find attractive and appealing. Sophistication often brings with it a peculiar disdain for popular appeal, where the very aspects of a design that make it appeal to many people distress some intellectuals.

There is one other distinction among the levels: time. The visceral and behavioral levels are about "now," your feelings and experiences while actually seeing or using the product. But the reflective level extends much longer—through reflection you remember the past and contemplate the future. Reflective design, therefore, is about long-term relations, about the feelings of satisfaction produced by owning, displaying, and using a product. A person's self-identity is located within the reflective level, and here is where the interaction between the product and your identity is important as demonstrated in pride (or shame) of ownership or use. Customer interaction and service matter at this level.

Working with the Three Levels

The ways in which the three levels interact are complex. Still, for purposes of application, it is possible to make some very useful simplifications. So, although the scientist in me protests that what I am about to say is far too simple, the practical, engineering, designer side of me says that the simplification is good enough, and, more important, useful.

The three levels can be mapped to product characteristics like this:

Visceral design	›	Appearance
Behavioral design	›	The pleasure and effectiveness of use
Reflective design	›	Self-image, personal satisfaction, memories

Even these simplifications are difficult to apply. Should some products be primarily visceral in appeal, others behavioral, others reflective? How does one trade off the requirements at one level against those of the others? How do visceral pleasures translate into products? Won't the same things that excite one group of people dismay others? Similarly, for the reflective level, wouldn't a deep reflective component be attractive to some and bore or repel others? And, yes, we can all agree that behavioral design is important—nobody is ever against usability—but just how much in the total scheme of things? How does each of the three levels compare in importance with the others?

The answer is, of course, that no single product can hope to satisfy everyone. The designer must know the audience for whom the product is intended. Although I have described the three levels separately, any real experience involves all three: a single level is rare in practice, and if it exists at all is most likely to come from the reflective level than from the behavioral or the visceral.

Consider the visceral level of design. On the one hand, this would appear to be the easiest level to appeal to since its responses are biological and similar for everyone across the world. This does not necessarily translate directly into preferences. Furthermore, although all people have roughly the same body shape, the same number of limbs, and the same mental apparatus, in detail, they differ considerably. People are athletic or not, energetic or lazy. Personality theorists divide people along such dimensions as extraversion, agreeableness, conscientiousness, emotional stability, and openness. To designers, this means that no single design will satisfy everyone.

In addition, there are large individual differences in the degree of a visceral response. Thus, while some people love sweets and especially chocolate (some claim to be addicts or "chocoholics"), many can ignore them, even if they like them. Almost everyone initially dislikes bitter and sour tastes, but you can learn affection for them, and they are often the components of the most expensive meals. Many foods loved by adults were intensely disliked at first taste: coffee, tea, alcoholic drinks, hot pepper, and even foods—oysters, octopus, and eyeballs—that make many people squeamish. And although the visceral system has evolved to protect the body against danger, many of our most popular and sought-after experiences involve horror and danger: horror novels and movies, death-defying rides, and thrilling, risky sports. And, as I have already mentioned, the pleasure of risk and perceived danger varies greatly among people. Such individual differences are the basic components of personality, the distinctions among people that make each of us unique.

Go outside. Get some air.
Watch a sunset.
Boy, does that get old fast.

—XBOX advertisement
(Microsoft's video game player)

THE TEXT of Microsoft's ad campaign for XBOX appeals to teens and young adults (whatever their actual age) who seek fast, exciting games with high visceral arousal, contrasting these people with those who prefer the commonly accepted norm that sunsets and fresh air are emotionally satisfying. The advertisement pits the reflective emotions of being outside and sitting quietly, enjoying the sunset against the continuous visceral and behavioral thrill of the fast-moving, engaging video game. Some people can spend hours watching sunsets. Some get bored after the first few seconds: "Been there, done that," is the refrain.

With the large range of individual, cultural, and physical differences among the people of the world, it is impossible for a single product to satisfy everyone. Some products are indeed marketed to everyone across the world, but they can succeed only if there are no real alternatives or if they do manage to reposition their appeal to different people through the adroit use of marketing and advertising. Hence, Coca-Cola and Pepsi-Cola manage worldwide success, in part capitalizing on a universal liking for sweet beverages, in part through sophisticated, culture-specific advertising. Personal computers are successful throughout the world because their benefits overcome their (numerous) deficiencies, and because there really is no choice. But most products have to be sensitive to the differences among people.

The only way to satisfy a wide variety of needs and preferences is to have a wide variety of products. Many product categories specialize, each catering to a different audience. Magazines are a good example. The world has tens of thousands of magazines (almost 20,000 in the United States alone). It is the rare magazine that tries to cater to everyone. Some magazines even flaunt their specialness, pointing out that they aren't for everyone, just for the people who match a particular set of interests and style.

Most product categories—home appliances, shop or gardening tools, furniture, stationery goods, automobiles—are manufactured and distributed differently across the world, with a wide variety of styles and form depending upon the needs and preferences of the market segment for whom they are targeted. Market segmentation is the marketing phrase used for this approach. Automobile companies bring out a variety of models, and different companies often emphasize different market segments. Some are for older, more sedate established people, some for the young and adventurous. Some are for those who truly need to go into the wilderness and travel through rivers and forests, up and down steep inclines, through mud, sand, and snow. Others are for those who like the reflective image of appearing to do such adventurous activities, but who will never actually do them.

Another important dimension for a product is its appropriateness to setting. In some sense, this point applies to all of human behavior:

What is appropriate and indeed preferred in one setting may be most inappropriate and rejected in another. All of us have learned to modulate our language, speaking differently when in casual interaction with our friends than when in formal presentation at a serious business meeting, or with the parents of our friends, or with our professors. Clothes that are appropriate for late-night clubs are inappropriate in business. A product that is cute and snuggly, or that conveys a humorous, playful image, is probably not appropriate for the business setting. Similarly, an industrial-style design, appropriate for the factory floor, would not be for the home kitchen or living room.

Computers sold to the home marketplace often are more powerful and have better sound systems than computers used in business. In fact, many business computers do not have some of the standard features of home machines, such as dial-out modems, sound systems, or DVD players. The reason is that these aspects of the machine are necessary for entertainment or game playing, activities not appropriate in the serious world of business. If a computer looks too attractive and playful, management may reject it. Some people feel that this hurt the sales of Apple's Macintosh computer. The Macintosh is considered a home, education, or graphics machine, not appropriate for business workers. This is an image problem because in fact, computers are pretty much the same, whether made by Apple or some other manufacturer, whether running the Macintosh or the Windows operating system, but images and psychological perceptions determine what people will buy.

The distinction between the terms *needs* and *wants* is a traditional way of describing the difference between what is truly necessary for a person's activities (needs) versus what a person asks for (wants). Needs are determined by the task: A pail is needed to carry water; some sort of carrying case is needed to transport papers back and forth to work. Wants are determined by culture, by advertising, by the way one views oneself and one's self-image. Although a student's backpack or even a paper bag would work perfectly fine for carrying papers, it might be embarrassing to carry one into a serious "power" business meeting. Embarrassment is, of course, an emotion that reflects one's sense of the appropriateness of behavior and is really all in the mind. Product designers and marketing executives know that wants can often be more powerful than needs in determining the success of a product.

Satisfying people's true needs, including the requirements of different cultures, age groups, social and national requirements, is difficult. Now add the necessity to cater to the many wants—whims, opinions, and biases—of the people who actually purchase products, and the task becomes a major challenge. Note that many people purchase products for others, whether it be the purchasing department of a company trying to minimize cost, a parent buying for a child, or a contractor equipping a home with appliances that might enhance the sale of a house, whether or not the occupants would ever use them. To some designers, the challenge seems overwhelming. To others, it inspires.

One example of the challenge comes from the marketing of consoles for playing video games. Video game machines are clearly aimed squarely at the traditional game market: young males who love excitement and violence, who want rich graphics and sound and have quick reflexes, whether for sports or first-person shooting matches. The design of the machine reflects this image, as does the advertising: big, hefty, powerful, technical; young, virile, male. For this market, the game machines have been so wildly successful that the sales of video games exceeds the box-office sales of movies.

But although the design of these machines still seems to be targeted at young males, the actual market for video games is much broader. The average age is now around thirty, roughly as many women as men play, and the appeal is worldwide. In the United States, roughly half the population plays some sort of video game. Many of these games are no longer wild and violent. I discuss video games as a new genre of entertainment and literature in chapter 4, but here I want to focus on the fact that, despite the broader audience, the physical design of the game consoles has not been changed to meet the growing popularity. The design focus upon young, excitable males limits the potential sales to a fraction of the possible audience, excluding not only many girls and women but also many men. This rich potential is completely untapped.

Moreover, the potential uses of video games extends far beyond the playing of games. They could be excellent teaching devices. In playing a game, you have to learn an amazing variety of skills and knowledge. You attend deeply and seriously for hours, weeks, even months. You read books and study the game thoroughly, doing active problem solving and working with other people. These are precisely the activities of an effective learner, so what marvelous learning could be experienced if only we could use this same intensity when interacting with meaningful topics. Thus, game machines have huge potential for everyone, but it has not been systematically addressed.

To break out of the traditional video game market, the industry needs to project a different kind of appeal. Here is where the three levels of design come into play. At the visceral level, the physical appearances of the consoles and controllers need to be changed. Different markets should have different designs. Some designs should reflect a warmer, more feminine approach. Some should look more serious, more professional. Some should have a more reflective appeal, especially for the educational marketplace. These changes wouldn't make the product dull and unexciting. They could make it as inviting and attractive as before, but emphasize different aspects of its potential. Its appearance should match its usage and audience.

Today, the behavioral design of many games revolves around powerful graphics and fast reflexes. Skill at operating the controls is one of the features distinguishing the beginning from the advanced player. To branch out into other arenas requires changing the behavioral characteristics so that they emphasize rich, detailed graphics and informative structures. In many domains, the emphasis should be on content, not on the skill of using the device, so ease of use should be stressed. Where content matters, the user should not have to spend time mastering the device, but rather should be able to devote time and effort toward mastering the content, enjoying the presentations, and exploring the domain.

The reflective design of today's games projects a product image that is consistent with the sleek powerful appearance of the console and the fast reflexes required of the player. This has to be changed. Advertisements should promote the device as a learning and educational tool for people of all ages. One form of console should continue to project the image of powerful game machine. Others should be positioned to be an intelligent guide to activities such as cooking or auto mechanics or woodworking. And others should be positioned as an aid to learning. Each with different appearances, different modes of operation, and different advertising and marketing messages.

Now imagine the outcome. The device that used to be specialized for the playing of video games takes on different appearances, depending upon its intended function. In the garage, the device would look like shop machinery, with a serious, rugged appearance, impervious to damage. It would serve as tutor and assistant, displaying automobile manuals, mechanical drawings, and short videos of the required steps to maintain or upgrade the auto. In the kitchen, it matches the décor of kitchen appliances and becomes a cooking aid and tutor. In the living room, it fits with the furniture and books and becomes a reference manual, perhaps an encyclopedia, tutor, and player of reflective games (such as go, chess, cards, word games). And for the student, it is a source of simulations, experiments, and extensive exploration of interesting, well-motivated topics, but topics carefully chosen so that, in the process of enjoying the adventure, you automatically learn the fundamentals of your field. Designs appropriate to the audience, the location, and the purpose. Everything I have described here is doable. It simply hasn't yet been done.

• • •

Objects That Evoke Memories

True, long-lasting emotional feelings take time to develop: they come from sustained interaction. What do people love and cherish, despise and detest? Surface appearance and behavioral utility play relatively minor roles. Instead, what matters is the history of interaction, the associations that people have with the objects, and the memories they evoke.

Consider keepsakes and mementoes, postcards and souvenir monuments, such as the model of the Eiffel Tower shown in figure 2.2. These are seldom considered beautiful, seldom thought of as works of art. In the world of art and design they are called *kitsch*. This term of derision for the cheap and vulgar has been applied, says the *Columbia Electronic Encyclopedia*, "since the early 20th century to works considered pretentious and tasteless. Exploitative commercial objects such as *Mona Lisa* scarves and abominable plaster reproductions of sculptural masterpieces are described as kitsch, as are works that claim artistic value but are weak, cheap, or sentimental." "Sentimental" means, according to the *American Heritage Dictionary*, "resulting from or colored by emotion rather than reason or realism." "Emotion rather than reason"—well, yes, that is precisely the point.

Yogi Berra put it this way: "Nobody goes there anymore. It's too crowded." Or, translating this to design, "Nobody likes kitsch, it's too popular." Yup. If too many people like something, there must be something wrong with it. But isn't that very popularity telling us something? We should stop to consider just why it is popular. People find value in it. It satisfies some basic need. Those who deride kitsch are looking at the wrong aspects.

Yes, the cheap reproductions of famous paintings, buildings, and monuments are "cheap." They have little artistic merit, being copies of existing work, and poor copies at that. There is little intellectual depth, for the creativity and insight is part of the original, not the copy. Similarly, most souvenirs and popular trinkets are gaudy, schmaltzy, "excessively or insincerely emotional." But while this may be true of the object itself, that object is important only as a symbol, as a source of memory, of associations. The word *souvenir* means "a token of remembrance, a memento." The very sentimentality the world of art or design derides is the source of something's strength and popularity. Kitschy objects of the sort shown in figure 2.2 do not pretend to be art—they are aids to memory.

FIGURE 2.2
A souvenir monument.
Although often denounced as "kitsch," unworthy of being considered as art, souvenirs are rich in emotional meanings because of the memories they evoke. *(Author's collection.)*

In the world of design, we tend to associate emotion with beauty. We build attractive things, cute things, colorful things. However important these attributes, they are not what drive people in their everyday lives. We like attractive things because of the way they make us feel. And in the realm of feelings, it is just as reasonable to become attached to and love things that are ugly as it is to dislike things that would be called attractive. Emotions reflect our personal experiences, associations, and memories.

In *The Meaning of Things*, a book that should be required reading for designers, Mihaly Csikszentmihalyi and Eugene Rochberg-Halton study what makes things special. The authors went into homes and interviewed the residents, trying to understand their relationship to the things about them, to their material possessions. In particular, they asked each person to show things that were "special" to him or her, and then, in the extensive interviews, explored what factors made them so. Special objects turned out to be those with special memories or associations, those that helped evoke a special feeling in their owners. Special items all evoked stories. Seldom was the focus upon the item itself: what mattered was the story, an occasion recalled. Thus, one woman interviewed by Csikszentmihalyi and Rochberg-Halton pointed to her living-room chairs and said: "They are the first two chairs me and my husband ever bought, and we sit in them and I just associate them with my home and having babies and sitting in the chairs with babies."

We become attached to things if they have a significant personal association, if they bring to mind pleasant, comforting moments. Perhaps more significant, however, is our attachment to places: favorite corners of our homes, favorite locations, favorite views. Our attachment is really not to the thing, it is to the relationship, to the meanings and feelings the thing represents. Csikszentmihalyi and Rochberg-Halton identify "psychic energy" as the key. Psychic energy, by which we mean mental energy, mental attention. Csikszentmihalyi's concept of "flow" provides a good example. In the flow state, you become so engrossed and captured by the activity being performed that it is as if you and the activity were one: You are in a trance where the world disappears from consciousness. Time stops. You are only aware of the activity itself. Flow is a motivating, captivating, addictive state. It can arise from transactions with valued things. "Household objects," say Csikszentmihalyi and Rochberg-Halton, "facilitate flow experiences in two different ways. On the one hand, by providing a familiar symbolic context they reaffirm the identity of the owner. On the other hand, objects in the household might provide opportunities for flow directly, by engaging the attention of people."

Perhaps the objects that are the most intimate and direct are those that we construct ourselves, hence the popularity of home-made crafts, furniture, and art. Similarly, personal photographs, even though they may be technically inferior: blurred, heads cut off, or fingers obscuring the image. Some may have faded, or be torn and repaired with tape. Their surface appearance is less important than their ability to evoke the memory of particular people and events.

This point was vividly dramatized for me in 2002 when I walked through the exhibits on display at the San Francisco Airport. This is one of the world's most interesting museums—especially for people like me who are fascinated by everyday things, by the impact of technology upon people and society. This exhibition, "Miniature Monuments," was about the role of souvenirs in evoking memory. The show displayed hundreds of miniature monuments, buildings, and other souvenirs. The items were not on display for their artistic quality, but to applaud their sentimental value, for the memories they evoked and, in brief, for their emotional impact upon their owners. The text that accompanied the exhibition described the role of souvenir monuments thusly:

> The marvel of souvenir buildings is that the identical miniature sparks in each of us extravagantly different webs of remembrance.
>
> While the purpose of all monuments is to cause us to remember, their subjects have a wide range. Great people and important events; wars and their casualties; and the history of Astoria, Oregon, are memorialized in the monuments represented in miniature.
>
> These souvenirs serve two purposes, though. Even as a copper-plated pot metal replica of Lincoln's Tomb in Springfield, Illinois, causes us to remember the sixteenth president, it also prods recollection of the monument itself. Monuments may remember significant people and events; architectural miniatures remember the monuments.
>
> The architect Bruce Goff has remarked, "In architecture, there's the reason you do something, and then there's the real reason." With souvenir buildings, despite their ostensible (if purposeless) functions, their real reason remains the provocation of human memory.

Those of us viewing these miniatures did not necessarily have any emotional attachment to the objects—after all, they weren't ours; they were collected and displayed by someone else. Still, as I strolled around, I was most attracted to souvenirs of places I had myself visited, perhaps because they brought back memories of those visits. Had any one been emotionally negative, however, I would have quickly moved past it to escape—not the object but the memories it called forth in me.

PHOTOGRAPHS, MORE than almost anything else, have a special emotional appeal: they are personal, they tell stories. The power of personal photography lies in its ability to transport the viewer back in time to some socially relevant event. Personal photographs are mementos, reminders, and social instruments, allowing memories to be shared across time, place, and people. In the year 2000, there were about 200 million cameras in the United States alone, or around two cameras per household; with these cameras people took around 20 billion photographs. With the advent of digital cameras, it is no longer possible to know just how many pictures are being taken, but probably a lot more.

Although pictures are loved for the memories they maintain, the technologies of digital picture transmission, printing, file sharing, and display are sufficiently complex and time-consuming as to prevent many people from saving, retrieving, and sharing the pictures they cherish.

Numerous studies have shown that the work required to transform a picture in the camera into a print that can be shared defeats many people. Thus, while lots of pictures are taken, not all the film gets developed. Of the film that is developed, some of it is never looked at. Of the pictures that are looked at, many are simply put back into the envelope and then filed away in a box, never to be looked at again. (People in the photography industry call these "shoe boxes," because the storage is often within the cardboard boxes in which shoes come.) Some people carefully arrange their pictures in photo albums, but many of us have unused photo albums stored in closets or bookcases.

One of the most precious resources of the modern household is time, and the effort to take care of all those wonderful photographs defeats their value. Even though taking photographs out of an envelope and organizing them in photograph albums is about as simple a way of doing this job as can be imagined, most people don't do it. I don't.

Digital cameras change the emphasis, but not the principle. It is relatively easy to take digital photographs, easy to share them from the display on the camera itself. It is more difficult to print the pictures or email them to friends and acquaintances. Despite the power of the personal computer, paper prints of photographs are easier to take care of and display than are electronic versions. With electronic pictures comes the problem of storing them in some way that you can find them again later.

Thus, although we like to look at photographs, we do not like to take the time to do the work required to maintain them and keep them accessible. The design challenge is to keep the virtues while removing the barriers: make it easier to store, send, share. Make it easier to find just the desired pictures years after they have been taken and put into storage. These are not easy problems, but until they are overcome, we will not reap the full benefits of photography.

Portraits of family, though, are different. Wander through many places of work, and you'll see on desk, bookcase, and walls framed photographs of a person's family: husband, wife, son, daughter—family portraits, family snapshots—and occasionally parents. Yes, there are also ceremonial pictures of the person with the company president or other dignitaries, pictures of awards, and, in academic offices, conference photographs, where all the participants have gathered together sometime during the conference for the ritual photograph, which ends up published in the conference proceedings and posted on walls.

But, I hasten to add, this personal display is very culture-sensitive. Not all cultures display such personal symbols. In some countries, the display of personal photographs in the office is extremely rare, and in the home it can be infrequent. Instead, visitors are shown the photograph album, with each photograph lovingly pointed at and described. Some cultures prohibit photographs altogether. Still, the major nations of the world on all continents take billions of photographs, so that even if they are not on public display, they serve a powerful emotional role.

Photographs are clearly important to people's emotional lives. People have been known to rush back into burning homes to save treasured photographs. Their comforting presence maintains family bonds even when the people are separated. They assure permanence of the memories and are often passed from generation to generation. In the days before photography, people hired portrait painters to create images of loved or respected ones. The task required long sittings and produced more formal results. Painting had the virtue that the artist could change people's appearance to fit their desires rather than be restricted to the reality of the photograph. (Nowadays, with digital tools readily available, photographs, too, are easily doctored. I plead guilty to altering a family group photograph, replacing the scowling face of one family member with a happy, smiling face from a photograph of that person at a different occasion. Nobody has ever noticed the modification, not even the person who was modified.) Today, even with the ubiquity of personal cameras, portrait photographers maintain a lively business, in part because only professionals usually have the skills required for lighting and framing the shot so as to produce a high-quality picture.

Photographs can bring back only sights, not sounds. David Frohlich, a research scientist at the Hewlett Packard Laboratories in Bristol, England, has been developing a system he calls "audiophotography," photographs that combine an audio track, capturing the sounds on the scene surrounding the instant when the picture was taken. (Yes, the recording can start before the photograph is taken, one of the magical possibilities of modern technology.) Amy Cowen, who wrote about Frohlich's work, described its importance this way: "With every photo there is a story, a moment, a memory. As time passes, however, the user's ability to recall the details needed to evoke the moment the picture records fades. Adding sound to a photo can help keep the memories intact."

Frohlich points out that today's technology allows us both to capture the sounds occurring around the time a photograph is being taken and also to play them back while it is being shown in an album. The sounds capture the emotional setting in a far richer way than can the image itself. Imagine a family group photograph where, in the twenty seconds prior to the taking of the picture, the voices of family members joshing among themselves ("Mary, stop scowling" and "Henry, quick, stand between Frank and Uncle Oscar") are also recorded—possibly followed by giggling and relief in the twenty seconds after the photo was taken. Frohlich describes the possibilities this way: "Ambient sounds recorded around the moment of image capture provide an atmosphere or mood that can really help you remember the original event better. Nostalgic music set to a photo can evoke more feelings and memories of the era in which the photo was taken, and a spoken story can help others to interpret the meaning of the photo, especially in the absence of the photographer."

Feelings of Self

Memories reflect our life experiences. They remind us of families and friends, of experiences and accomplishments. They also serve to reinforce how we view ourselves. Our self-image plays a more important role in our lives than we like to admit. Even those who deny any interest in how others view them actually do care, if only by making sure that everyone else understands that they don't. The way we dress and behave, the material objects we possess, jewelry and watches, cars and homes, all are public expressions of our selves.

The concept of self appears to be a fundamental human attribute. It is difficult to imagine how it could be otherwise, given what we know of the mechanisms of mind and the roles that consciousness and emotion play. The concept is deeply rooted in the reflective level of the brain and highly dependent upon cultural norms. It is, therefore, difficult to deal with in design.

In psychology, the study of the self has become a big industry, with books, societies, journals, and conferences. But "self" is a complex concept: It is culturally specific. Thus, Eastern and Western notions of self vary considerably, with the West placing more emphasis on the individual, the East on the group. Americans tend to want to excel as individuals, whereas Japanese wish to be good members of their groups and for others to be satisfied with their contributions. But even these characterizations are too broad and oversimplified. In fact, on the whole, people behave very similarly, given the same situation. It is culture that presents us with different situations. Thus, Asian cultures are more likely to establish a sharing, group attitude than are the cultures of Europe and the Americas, where individualistic situations are more common. But put Asians in an individualistic situation and Europeans or Americans in a social, sharing situation, and their behaviors are remarkably similar.

Some aspects of self seem to be universal, such as the desire to be well-thought-of by others, even if the behavior others praise differs across cultures. This desire holds both in the most individualistic societies, which admire deviance, and in the most group-oriented societies, which admire conformance.

The importance of other people's opinions is, of course, well known to the advertising industry, which tries to promote products through association. Take any product and show it alongside happy, contented people. Show people doing things that an intended purchaser is likely to fantasize about, such as romantic vacations, skiing, exotic locations, eating in foreign lands. Show famous people, people who serve as role models or heroes to the customers, to induce in them, through association, a sense of worthiness. Products can be designed to enhance these aspects. In clothing fashion, one can have clothes that are neat and trim or baggy and nondescript, each deliberately inducing a different image of self. When company or brand logos are imprinted on clothes, luggage, or other objects, the mere appearance of the name speaks to others about your sense of values. The styles of objects you choose to buy and display often reflect public opinion as much as behavioral or visceral elements. Your choice of products, or where and how you live, travel, and behave are often powerful statements of self, whether intended or not, conscious or subconscious. For some, this external manifestation compensates for an internal, personal lack of self-esteem. Whether you admit it or not, approve or disapprove, the products you buy and your lifestyle both reflect and establish your self-image, as well as the images others have of you.

One of the more powerful ways to induce a positive sense of self is through a personal sense of accomplishment. This is one aspect of a hobby, where people can create things that are uniquely theirs, and, through hobby clubs and groups, share their achievements.

From the late 1940s through the mid-1980s the Heathkit Company sold electronic kits for the home handyperson. Build your own radio, your own audio system, your own television set. The people who constructed the kits felt immense pride in their accomplishments as well as a common bond with other kit builders. Putting together a kit was a personal feat: the less skilled the kit builder, the more that special feeling. Electronic experts did not take such pride in their kits; it was those who ventured forth without the expertise who felt so satisfied. Heathkit did an excellent job of aiding the first-time builder with what, in my opinion, were the best instruction manuals ever written. Mind you, the kits were not much less expensive than equivalent commercial electronic devices. People bought the kits for their high quality and for the feeling of accomplishment, not to save money.

In the early 1950s, the Betty Crocker Company introduced a cake mix so that people could readily make excellent tasting cakes at home. No muss, no fuss: just add water, mix, and bake. The product failed, even though taste tests confirmed that people liked the result. Why? An after-the-fact effort was made to find the reasons. As the market researchers Bonnie Goebert and Herma Rosenthal put it: "The cake mix was a little too simple. The consumer felt no sense of accomplishment, no involvement with the product. It made her feel useless, especially if somewhere her aproned mom was still whipping up cakes from scratch."

Yes, it was too easy to make the cake. Betty Crocker solved the problem by requiring the cook to add an egg to the mix, thereby putting pride back into the activity. Clearly, adding an egg to a prepared cake mix is not at all equivalent to baking a cake "from scratch" by using individual ingredients. Nonetheless, adding the egg gave the act of baking a sense of accomplishment, whereas just mixing water into the cake mix seemed too little, too artificial. Goebert and Rosenthal summarized the situation: "The real problem had nothing to do with the product's intrinsic value, but instead represented the emotional connection that links a product to its user." Yes, it's all about emotion, about pride, about the feeling of accomplishment, even in making a cake from a prepared mix.

The Personality of Products

As we have seen, a product can have a personality. So, too, can companies and brands. Consider my proposed transformation of the video game device discussed earlier in this chapter. In one version, the machine would be a fast, powerful tool for exciting, visceral experiences: loud booming sounds and fast-paced adventure. In a different version, it would be a cooking assistant: animated, but informative, with menus for meals and videos that demonstrate just how to prepare the food. In still another, it might be calm, but authoritative, guiding repair work on an automobile or construction of woodworking projects.

In each manifestation, that product's personality would change. The product would look and behave differently in the different settings appropriate to use and target audience. The style of behavioral interaction could differ: filled with slang and informal language in the game setting; polite and formal for the kitchen. But like human personality, once established, all aspects of a design must support the intended personality structure. A mature cooking tutor should not suddenly start spouting obscenities. A shop assistant should probably not discuss the philosophical implications of quality in automobile

Figure 2.3
Fashion from the seventeenth century.
On the left, Maria Anna of Bavaria, crown princess of France. On the right, a "young elegant."
(Braun et al., courtesy of Northwestern University Library.)

design, quoting from R. M. Pirsig's *Zen and the Art of Motorcycle Maintenance* whenever a repair is attempted.

Personality is, of course, a complex topic in its own right. A simplified way of thinking of product personality is that it reflects the many decisions about how a product looks, behaves, and is positioned throughout its marketing and advertisements. Thus, all three levels of design play a role. Personality must be matched to market segment. And it must be consistent. Think about it. If a person or product has an obnoxious personality, at least you know what to expect: you can plan for it. When behavior is inconsistent and erratic, it is difficult to know what to expect, and occasional positive surprises are not enough to overcome the frustration and irritation caused by never quite knowing what to expect.

The personalities of products, companies, and brands need as much tending to as the product itself.

The *American Heritage Dictionary* defines *fashion, style, mode,* and *vogue* thus: "These nouns refer to a prevailing or preferred manner of dress, adornment, behavior, or way of life at a given time. Fashion, the broadest term, usually refers to what accords with conventions adopted by polite society or by any culture or subculture: a time when long hair was the fashion. Style is sometimes used interchangeably with fashion, but like mode often stresses adherence to standards of elegance: traveling in style; miniskirts that were the mode in the late sixties. Vogue is applied to fashion that prevails widely and often suggests enthusiastic but short-lived acceptance: a video game that was in vogue a few years ago."

The very existence of the terms *fashion*, *style*, *mode*, and *vogue* demonstrates the fragility of the reflective side of design. What is liked today may not be tomorrow. Indeed, the reason for the change is the very fact that something was once liked: When too many people like something, then it is no longer deemed appropriate for the leaders of a society to partake of it. After all, goes the thinking, how can one be a leader unless one is different, doing today what others will do tomorrow, and doing tomorrow what they will be doing after that? Even the rebellious have to change continually, carefully noticing what is in fashion so as not to be following it, carefully creating their own fashion of counterfashion.

How does a designer cope with popular taste if it has little to do with substance? Well, it depends upon the nature of the product and the intentions of the company producing it. If the product is something fundamental to life and well-being, then the proper response is to ignore continual shifts in popular sentiment and aim for long-lasting value. Yes, the product must be attractive. Yes, it should be pleasurable and fun. But it must also be effective, understandable, and appropriately priced. In other words, it must strive for balance among the three levels of design.

In the long run, simple style with quality construction and effective performance still wins. So a business that manufactures office machines, or basic home appliances, or web sites for shipping, commerce, or information, would be wise to stick to the fundamentals. In these cases, the task dictates the design: make the design fit the task, and the product works more smoothly and is bound to be more effective across a wide range of users and uses. Here is where the number of different products is determined by the nature of particular tasks and the economics.

There is a set of products, however, whose goals are entertainment, or style, or perhaps enhancement of a person's image. Here is where fashion comes into play. Here is where the huge individual differences in people and cultures are important. Here the person and market segment dictate the design. Make the design appropriate to the market segment that forms the target audience. It is probably necessary to have multiple versions of the design for different market segments. And it is probably necessary to do rapid changes in style and appearance as the market dictates.

Designing for the whims of fashion is tricky. Some designers may see it as a difficult challenge, others, as an opportunity. In some sense, the division often breaks between large and small companies, or between market leaders and the competition. To the market leader, the continual changes in people's fashion, and the wide variety of preferences for the same product across the world, are huge challenges. How can the company ever keep up? How does it track all the changes and even anticipate them? How does it keep the many necessary product lines effective? To the competitive companies, however, the same issues represent an opportunity. Small companies can be nimble, moving rapidly into areas and using approaches that the more conservative larger companies hesitate to try. Small companies can be outrageous, different, and experimental. They can exploit the public's interests, even if the product is initially purchased by only a few. Large companies attempt to experiment by spinning off smaller, more nimble divisions, sometimes with unique names that make them appear to be independent of their parent. All in all, this is the ever-changing, continual battleground of the consumer marketplace, where fashion can be as important as substance.

IN THE world of products, a brand is an identifying mark, the symbol that represents a company and its products. Particular brands produce an emotional response that draws the consumer toward the product or away from it. Brands have taken on the emotional representation. They carry with them an emotional response that guides us toward a product or away from it. Sergio Zyman, former chief marketing officer of Coca-Cola, has said that "emotional branding is about building relationships; it is about giving a brand and a product long-term value." But it is more: it involves the entire relationship of the product to the individual. Again, in Zyman's words: "Emotional branding is based on that unique trust that is established with an audience. It elevates purchases based on need to the realm of desire. The commitment to a product or an institution, the pride we feel upon receiving a wonderful gift of a brand we love or having a positive shopping experience in an inspiring environment where someone knows our name or brings an unexpected gift of coffee—these feelings are at the core of Emotional Branding."

Some brands are simply informative, essentially naming a company or its product. But on the whole, the brand name is a symbol that represents one's entire experience with a product and the company that produces it. Some brands represent quality and high prices. Some represent a focus upon service. Some represent value for money. And some brands stand for shoddy products, for indifferent service, or for inconvenience at best. And, of course, most brand names are meaningless, carrying no emotional power at all.

Brands are all about emotions. And emotions are all about judgment. Brands are signifiers of our emotional responses, which is why they are so important in the world of commerce.

THIS CONCLUDES part I of the book: the basic tools of emotional design. Attractive things do work better—their attractiveness produces positive emotions, causing mental processes to be more creative, more tolerant of minor difficulties. The three levels of processing lead to three corresponding forms of design: visceral, behavioral, and reflective. Each plays a critical role in human behavior, each an equally critical role in the design, marketing, and use of products. Now it is time to explore how this knowledge is put to work.

Part Two

Design in Practice

FIGURE 3.1

Water bottles. The ones on the left and the right are clearly aimed to please at the visceral level; the middle one, well, it is efficient, it is inexpensive, and it works. The bottle on the left, for *Perrier* water, has become so well known that the shape and its green color are the brand. The bottle on the right is by *TyNant,* a bottle of such a pleasant shape coupled with its deep, cobalt blue color that people save the empty ones to use as vases. The clear plastic bottle is by *Crystal Geyser:* simple, utilitarian, effective when you need to carry water with you.

(Author's collection.)

CHAPTER THREE

Three Levels of Design: Visceral, Behavioral, and Reflective

I remember deciding to buy Apollinaris, a German mineral water, simply because I thought it would look so good on my shelves. As it turned out, it was a very good water. But I think I would have bought it even though it was not all that great.

The nice interplay between the bottle's green and the label's beige and red as well as the font used for the brand turned this product of mass consumption into a decoration accessory for your kitchen.

—*Hugues Belanger*
email, 2002

IT WAS LUNCHTIME. My friends and I were in downtown Chicago, and we decided to try Café des Architectes in the Sofitel Hotel. As we entered the bar area, a beautiful display greeted us: water bottles, the sort you can buy in a food market, set out as works of art.

The entire rear wall of the bar was like an art gallery: frosted glass, subtly lit from behind, from floor to ceiling; shelves in front of the glass, each shelf dedicated to a different type of water. Blue, green, amber—all the wonderful hues, the glass gracefully illuminating them from behind, shaping the play of color. Water bottles as art. I resolved to find out more about this phenomenon. How did the packaging of water become an art form?

"Walk down a grocery aisle in any town in the U.S., Canada, Europe, or Asia and there is a virtual tidal wave of bottled water brands," is how one web site that I consulted put it. Another web site emphasized the role of emotion: "Package designers and brand managers are looking beyond graphic elements or even the design as a whole to forge an emotional link between consumers and brands." The selling of premium bottled water in major cities of the world, where the tap water is perfectly healthful, has become a big business. Water sold in this way is more expensive than gasoline. Indeed, the cost is part of the attraction where the reflective side of the mind says, "If it is this expensive, it must be special."

And some of the bottles are special, sensuous, and colorful. People keep the empty bottles, sometimes refilling them with tap water, which, of course, demonstrates that the entire success of the product lies in its package, not its contents. Thus, like wine bottles, water bottles serve as decorative additions to rooms long after they have fulfilled their primary purpose. Witness another web site: "almost everyone who enjoys TyNant Natural Mineral Water admits to keeping one or two around the home or office as an ornament, vase or the like. Photographers positively delight in the bottles' photogenic appeal." (In figure 3.1, the bottle with the flower in it is TyNant.)

How does one brand of water distinguish itself from another? Packaging is one answer, distinctive packaging that, in the case of water, means bottle design. Glass, plastic, whatever the material, the design becomes the product. This is bottling that appeals to the powerful visceral level of emotion, that causes an immediate visceral reaction: "Wow, yes, I like it, I want it." It is, as one designer explained to me, the "wow" factor.

The reflective side of emotion is involved as well, for the saved bottles can serve as reminders of the occasion when the beverage was ordered or consumed. Because both wine and expensive water are sometimes purchased for special occasions, the bottles serve as mementos of those occasions, taking on a special emotional value, becoming meaningful objects, not because of the objects themselves, but because of the memories they produce, and, as I noted in chapter 2, memories can trigger the powerful, long-lasting emotions.

What are the design factors in play here, where pure appearance is the issue, beauty that is all on the surface? This is where those genetic, hard-wired biological processes play their role. Here the designs are apt to be "eye candy," as sweet to the eye as the taste of candy to the mouth. Yet just as sweet-tasting candy is empty of nutritional value, so, too, is appearance empty beneath the surface.

Human responses to the everyday things of the world are complex, determined by a wide variety of factors. Some of these are outside the person, controlled by designer and manufacturer, or by advertising and such things as brand image. And some come from within, from your own, private experiences. Each of the three levels of design—visceral, behavioral, and reflective—plays its part in shaping your experience. Each is as important as the others, but each requires a different approach by the designer.

Visceral Design

Visceral design is what nature does. We humans evolved to coexist in the environment of other humans, animals, plants, landscapes, weather, and other natural phenomena. As a result, we are exquisitely tuned to receive powerful emotional signals from the environment that get interpreted automatically at the visceral level. This is where the lists of features in chapter 1 came from. Thus, the colorful plumage on male birds was selectively enhanced through the evolutionary process to be maximally attractive to female birds—as, in turn, were the preferences of female birds so as to discriminate better among male plumages. It's an iterative, co-adaptive process, each animal adapting over many generations to serve the other. A similar process occurs between males and females of other species, between co-adaptive life forms across species, and even between animals and plants.

Fruits and flowers provide an excellent example of the co-evolution of plants and animals. Nature's evolutionary process made flowers to be attractive to birds and bees, the better to spread their pollen, and fruits to be attractive to primates and other animals, the better to spread their seeds. Fruits and flowers tend to be symmetrical, rounded, smooth, pleasant to the touch, and colorful. Flowers have pleasant odors, and most fruits taste sweet, the better to attract animals and people who will eat them and then spread the seeds, whether by spitting or defecation. In this co-evolution of design, the plants change so as to attract animals, while the animals change so as to become attracted to the plants and fruits. The human love of sweet tastes and smells and of bright, highly saturated colors probably derives from this co-evolution of mutual dependence between people and plants.

The human preference for faces and bodies that are symmetrical presumably reflects selection of the fittest; non-symmetrical bodies probably are the result of some deficiency in the genes or the maturation process. Humans select for size, color, and appearance, and what you are biologically disposed to think of as attractive derives from these considerations. Sure, culture plays a role, so that, for example, some cultures prefer fat people, others thin; but even within those cultures, there is agreement on what is and is not attractive, even if too thin or too fat for specific likes.

When we perceive something as "pretty," that judgment comes directly from the visceral level. In the world of design, "pretty" is generally frowned upon, denounced as petty, trite, or lacking depth and substance—but that is the designer's reflective level speaking (clearly trying to overcome an immediate visceral attraction). Because designers want their colleagues to recognize them as imaginative, creative, and deep, making something "pretty" or "cute" or "fun" is not well accepted. But there is a place in our lives for such things, even if they are simple.

You can find visceral design in advertising, folk art and crafts, and children's items. Thus, children's toys, clothes, and furniture will often reflect visceral principles: bright, highly saturated primary colors. Is this great art? No, but it is enjoyable.

Adult humans like to explore experiences far beyond the basic, biologically wired-in preferences. Thus, although bitter tastes are viscerally disliked (presumably because many poisons are bitter), adults have learned to eat and drink numerous bitter things, even to prefer them. This is an "acquired taste," so called because people have had to learn to overcome their natural inclination to dislike them. So, too, with crowded, busy spaces, or noisy ones, and discordant, nonharmonic music, sometimes with irregular beats: all things that are viscerally negative but that can be reflectively positive.

The principles underlying visceral design are wired in, consistent across people and cultures. If you design according to these rules, your design will always be attractive, even if somewhat simple. If you design for the sophisticated, for the reflective level, your design can readily become dated because this level is sensitive to cultural differences, trends in fashion, and continual fluctuation. Today's sophistication runs the risk of becoming tomorrow's discard. Great designs, like great art and literature, can break the rules and survive forever, but only a few are gifted enough to be great.

At the visceral level, physical features—look, feel, and sound—dominate. Thus, a master chef concentrates on presentation, arranging food artfully on the plate. Here good graphics, cleanliness, and beauty play a role. Make the car door feel firm and produce a pleasant chunking sound as it closes. Make the exhaust sound of the Harley Davidson motorcycle have a unique, powerful rumble. Make the body sleek, sexy, inviting, such as the classic 1961 Jaguar roadster of figure 3.2. Yes, we love sensuous curves, sleek surfaces, and solid, sturdy objects.

FIGURE 3.2

The 1961 Jaguar E-type: Viscerally exciting.

This automobile is a classic example of the power of visceral design: sleek, elegant, exciting. It is no surprise that the car is in the design collection of the New York Museum of Modern Art.

(Courtesy of the Ford Motor Corporation.)

Because visceral design is about initial reactions, it can be studied quite simply by putting people in front of a design and waiting for reactions. In the best of circumstances, the visceral reaction to appearance works so well that people take one look and say "I want it." Then they might ask, "What does it do?" And last, "And how much does it cost?" This is the reaction the visceral designer strives for, and it can work. Much of traditional market research involves this aspect of design.

Apple Computer found that when it introduced the colorful iMac computer, sales boomed, even though those fancy cabinets contained the very same hardware and software as Apple's other models, ones that were not selling particularly well. Automobile designers count on visual design to rescue a company. When Volkswagen reintroduced their classic "beetle" design in 1993, Audi developed the TT, and Chrysler brought out the PT Cruiser, sales for all three companies climbed. It's all in the appearance.

FIGURE 3.3
The sensual component of behavioral design.
Behavioral design emphasizes the use of objects, in this case, the sensual feel of the shower: a key, often overlooked component of good behavioral design. The Kohler WaterHaven Shower.
(Courtesy of the Kohler Co.)

Effective visceral design requires the skills of the visual and graphic artist and the industrial engineer. Shape and form matter. The physical feel and texture of the materials matter. Heft matters. Visceral design is all about immediate emotional impact. It has to feel good, look good. Sensuality and sexuality play roles. This is a major role of "point of presence" displays in stores, in brochures, in advertisements, and in other enticements that emphasize appearance. These may be a store's only chance of getting the customer, for many a product is purchased on looks alone. Similarly, otherwise highly rated products may be turned down if they do not appeal to the aesthetic sense of the potential buyer.

Behavioral Design

Behavioral design is all about use. Appearance doesn't really matter. Rationale doesn't matter. Performance does. This is the aspect of design that practitioners in the usability community focus upon. The principles of good behavioral design are well known and often told; indeed, I laid them out in my earlier book, *The Design of Everyday Things*. What matters here are four components of good behavioral design: function, understandability, usability, and physical feel. Sometimes the feel can be the major rationale behind the product. Consider the shower shown in figure 3.3. Imagine the sensual pleasure, the feel—quite literally—of the water streaming across the body.

IN MOST behavioral design, function comes first and foremost; what does a product do, what function does it perform? If the item doesn't do anything of interest, then who cares how well it works? Even if its only function is to look good, it had better succeed. Some well-designed items miss the target when it comes to fulfilling their purpose and thus deserve to fail. If a potato peeler doesn't actually peel potatoes, or a watch doesn't tell accurate time, then nothing else matters. So the very first behavioral test a product must pass is whether it fulfills needs.

On the face of it, getting the function right would seem like the easiest of the criteria to meet, but in fact, it is tricky. People's needs are not as obvious as might be thought. When a product category already exists, it is possible to watch people using the existing products to learn what improvements can be made. But what if the category does not even exist? How do you discover a need that nobody yet knows about? This is where the product breakthroughs come from.

Even with existing products, it is amazing how seldom the designers watch their customers. I visited a major software developer to meet with the design team for one of their more widely used products, one that has an overabundance of features, but nonetheless still fails to meet my everyday needs. I came prepared with a long list of problems that I had encountered while attempting to do routine activities. Moreover, I had checked with other dissatisfied users of this product. To my great surprise, much of what I told the design team seemed to be novel. "Very interesting," they kept saying, while taking copious notes. I was pleased that they paid attention to me, but disturbed by the fact that these rather basic points appeared to be new. Had they never watched people use their products? These designers—like many design teams in all industries—tend to keep to their desks, thinking up new ideas, testing them out on one another. As a result, they kept adding new features, but they had never studied just what patterns of activities their customers performed, just what tasks needed to be supported. Tasks and activities are not well supported by isolated features. They require attention to the sequence of actions, to the eventual goal—that is, to the true needs. The first step in good behavioral design is to understand just how people will use a product. This team had not done even this most elementary set of observations.

There are two kinds of product development: enhancement and innovation. Enhancement means to take some existing product or service and make it better. Innovation provides a completely new way of doing something, or a completely new thing to do, something that was not possible before. Of the two, enhancements are much easier.

Innovations are particularly difficult to assess. Before they were introduced, who would have thought we needed typewriters, personal computers, copying machines, or cell phones? Answer: Nobody. Today it is hard to imagine life without these items, but before they existed almost no one but an inventor could imagine what purpose they would serve, and quite often the inventors were wrong. Thomas Edison thought that the phonograph would eliminate the need for letters written on paper: business people would dictate their thoughts and mail the recordings. The personal computer was so misunderstood that several then-major computer manufacturers completely dismissed them: some of those once-large companies no longer exist. The telephone was thought to be an instrument for business, and in the early days, telephone companies tried to dissuade customers from using the phone for mere conversation and gossip.

One cannot evaluate an innovation by asking potential customers for their views. This requires people to imagine something they have no experience with. Their answers, historically, have been notoriously bad. People have said they would really like some products that then failed in the marketplace. Similarly, they have said they were simply not interested in products that went on to become huge market successes. The cellular telephone is a good example. It was originally thought to be of value to a limited number of business people. Very few people could imagine carrying one simply for personal interaction. Indeed, when individuals first purchased cell phones, they often explained that they didn't intend to use them, but that they were "in case I have an emergency." Predicting the popularity of a new product is almost impossible before the fact, even though it may seem obvious afterward.

Enhancements to a product come primarily by watching how people use what exists today, discovering difficulties, and then overcoming them. Even here, however, it can be more difficult to determine the real needs than might seem obvious. People find it difficult to articulate their real problems. Even if they are aware of a problem, they don't often think of it as a design issue. Ever struggle with a key, to discover that you are inserting it upside down? Or ever lock your keys inside the automobile? Or lock the car, only to realize that you left the windows open, so you have to unlock the car and lean inside to close them? In any of these cases, would you think these were design flaws? Probably not, probably you just blamed yourself. Well, they all could be corrected by appropriate designs. Why not design a symmetrical key that works no matter which way it is inserted into a lock? Why not design cars so that the key is required to lock the doors, making it much less likely that the car can be locked with the key inside? Why not make it possible to close the windows from outside the car? All of these designs now exist, but it took clever observations for the designers to recognize that the problems could be overcome.

Ever put batteries into a product in the wrong orientation? Why is this even possible? Why shouldn't batteries be designed so that they can only go into their slots in one orientation, making it impossible to insert them improperly? I suspect battery makers don't care, and that manufacturers who purchase and specify batteries for their equipment never considered that it was possible to do things better. Standard cylindrical batteries are excellent examples of poor behavioral design, of a failure to understand the problems facing people who must figure out just which orientation is required for each device—moreover in the face of warning labels that point out that the equipment might be damaged if the batteries are inserted incorrectly.

Consider the automobile. Sure, it is easy to note that storage areas ought to be bigger or seat adjustments easier, but how about such an obvious item as cup holders for beverages? People like to drink coffee and sodas while riding in their vehicles. Today this seems like an obvious necessity in an automobile, but it was not always thought so. Automobiles have been around roughly a century, but cup holders were not considered appropriate for their interiors until quite recently, and the innovation didn't come from the automobile manufacturers—they resisted them. What happened was that small manufacturers realized the need, probably because they had built cup holders for themselves, and then discovered that other people wanted them also. Soon, all sorts of add-on devices were being manufactured. These were relatively inexpensive and easy to install in a car: stick-on holders, magnetic holders, bean-bag holders. Some attached to the windows, some to the dashboards, and some to the space between the seats. It was only because these were so popular that manufacturers slowly started to add them as standard items inside the car. Now there is a vast array of clever cup holders. Some people claim that they purchased a particular automobile solely because of its cup holders. Buy a car solely because of the cup holders? Why not? If the car is used primarily for daily commuting and short errands around a city, convenience and comfort for drivers and passengers are the most important needs.

Even after the need for cup holders seemed obvious, German automobile manufacturers resisted them, explaining that automobiles were for driving, not drinking. (I suspect that this attitude reflects the old-fashioned German automobile design culture, which proclaims that the engineer knows best, and considers studies of real people driving their vehicles irrelevant. But if the automobile is only for driving, why do Germans provide ashtrays, cigarette lighters, and radios?) The Germans reconsidered only when decreases in sales in the United States were attributed to the lack of cup holders. Engineers and designers who believe they do not need to watch the people who use their products are a major source of the many poor designs that confront us.

My friends at the industrial design firm of Herbst LaZar Bell told me that they had been asked by a company to redesign their floor-cleaning machine to satisfy a long list of requirements. Cup holders were not on the list, but perhaps they should have been. When the designers visited maintenance workers in the middle of the night to observe just how they cleaned the floors of large commercial buildings, they discovered that workers had difficulty drinking coffee while manipulating the huge cleaning and waxing machines. As a result, the designers added cup holders. The new design had numerous major enhancements to the product in both appearance and behavior—visceral and behavioral design—and has proven to be a market success. How important was the cup holder to the success of the new design? Probably not much, except that it is symptomatic of the attention to true customer needs that signifies quality products. As Herbst LaZar Bell properly emphasizes, the real challenge to product design is "understanding end-user unmet and unarticulated needs." That's the design challenge—to discover real needs that even the people who need them cannot yet articulate.

How does one discover "unarticulated needs"? Certainly not by asking, not by focus groups, not by surveys or questionnaires. Who would have thought to mention the need for cup holders in a car, or on a stepladder, or on a cleaning machine? After all, coffee drinking doesn't seem to be a requirement for cleaning any more than for driving in an automobile. It is only after such enhancements are made that everyone believes them to be obvious and necessary. Because most people are unaware of their true needs, discovering them requires careful observations in their natural environment. The trained observer can often spot difficulties and solutions that even the person experiencing them does not consciously recognize. But once an issue has been pointed out, it is easy to tell when you have hit the target. The response of the people who actually use the product is apt to be something like, "Oh, yeah, you're right, that's a real pain. Can you solve that? That would be wonderful."

After function comes understanding. If you can't understand a product, you can't use it—at least not very well. Oh, sure, you could memorize the basic operating steps, but you probably will have to be reminded over and over again what they are. With a good understanding, once an operation is explained, you are apt to say, "Oh, yes, I see," and from then on require no further explanation or reminding. "Learn once, remember forever," ought to be the design mantra.

Without understanding, people have no idea what to do when things go wrong—and things always go wrong. The secret to good understanding is to establish a proper conceptual model. In *The Design of Everyday Things*, I pointed out that there are three different mental images of any object. First is the image in the head of the designer—call that the "designer's model." Then the image that the person using the device has of it and the way it works: call this the "user's model." In an ideal world, the designer's model and the user's model should be identical and, as a result, the user understands and uses the item properly. Alas, designers don't talk to the final users; they only specify the product. People form their models entirely from their observations of the product—from its appearance, how it operates, what feedback it provides, and perhaps, any accompanying written material, such as the advertising and manuals. (But most people don't read the manuals.) I called the image conveyed by the product and written material the "system image."

As Figure 3.4 indicates, designers can communicate with the eventual users only through the system image of a product. Thus, a good designer will make sure that the system image of the final design conveys the proper user model. The only way to find this out is through testing: develop early prototypes, then watch as people try to use

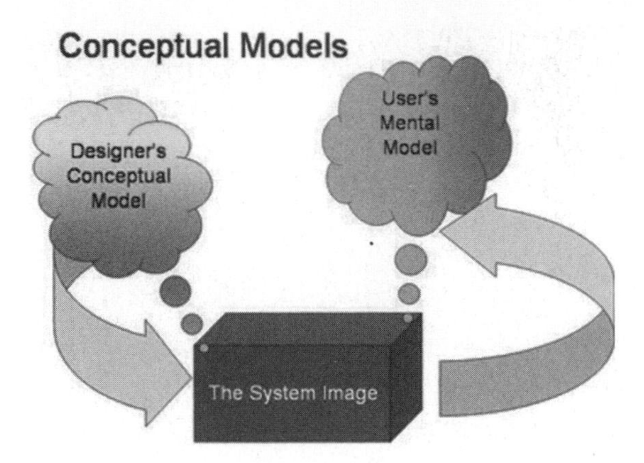

FIGURE 3.4

The designer's model, the system image, and the user's model. For someone to use a product successfully, they must have the same mental model (the user's model) as that of the designer (the designer's model). But the designer only talks to the user via the product itself, so the entire communication must take place through the "system image": the information conveyed by the physical product itself.

them. What is something with a good system image? Almost any design that makes apparent its operation. The rulers and margin setting in the word processor I use as I type this is one excellent example. The seat adjustment control shown in Figure 3.5 is another. Notice how the arrangement of the controls automatically refers to the operation each performs. Lift on the bottom seat control and the seat rises. Push forward on the vertical control and the seat back leans forward. That's good conceptual design.

An important component of understanding comes from feedback: a device has to give continual feedback so that a user knows that it is working, that any commands, button presses, or other requests have actually been received. This feedback can be as simple as the feel of the brake pedal when you depress it and the resultant slowing of the automobile, or a brief flash of light or sound when you push something. It is amazing, though, how many products still give inadequate

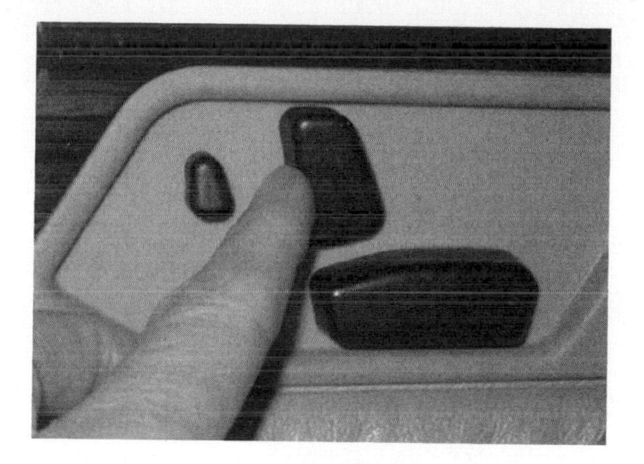

FIGURE 3.5

Seat controls—an excellent system image.

These seat controls explain themselves: the conceptual model is provided by the positioning of the controls to look just like the item being controlled. Want to change the seat adjustment? Push or pull, lift or depress the corresponding control and the corresponding part of the seat moves accordingly. *(Mercedes Benz seat controls; photograph by the author.)*

feedback. Most computer systems now display a clock face or an hourglass to indicate that they are responding, if slowly. If the delay is short, this indicator suffices, but it is completely inadequate if the delay lengthens. To be effective, feedback must enhance the conceptual model, indicating precisely what is happening and what yet remains to be done. Negative emotions kick in when there is a lack of understanding, when people feel frustrated and out of control—first uneasiness, then irritation, and, if the lack of control and understanding persists, even anger.

Usability is a complex topic. A product that does what is required, and is understandable, may still not be usable. Thus, guitars and violins do their assigned tasks well (that is, create music), they are quite simple to understand, but they are very difficult to use. The same is true of the piano, a deceptively simple-looking instrument. Musical instruments take years of dedicated practice to be used properly, and even then, errors and poor performance are common among nonprofessionals. The relative unusability of musical instruments is accepted, in part because we know of no other alternative, in part because the results are so worthwhile.

But most of the things you use in everyday life should not require years of dedicated practice. New items appear every week, but who has the time or energy to spend the time required to learn each one? Bad design is a frequent cause of error, often unfairly blamed on users rather than on designers. Errors can lead to accidents that not only are financially expensive but can cause injury or death. There is no excuse for such flaws, for we understand how to build functional, understandable, and usable things. Moreover, everyday things have to be used by a wide variety of people: short and tall, athletic and not, who speak and read different languages, who may be deaf or blind, or lack physical mobility or agility—or even hands. Younger people have different skills and abilities than older ones.

Usage is the critical test of a product: Here is where it stands alone, unsupported by advertising or merchandising material. All that matters is how well the product performs, how comfortable the person using it feels with the operation. A frustrated user is not a happy one, so it is at the behavioral stage of design that applying the principles of human-centered design pay off.

Universal design, designing for everyone, is a challenge, but one well worth the effort. Indeed, the "Universal Design" philosophy argues persuasively that designing for the handicapped, the hard of hearing or seeing, or those less agile than average invariably makes an object better for everyone. There is no excuse not to design usable products that everyone can use.

"HERE, TRY this." I am visiting IDEO, the industrial design company. I am being shown their "Tech Box," a big cabinet with an apparently endless set of small drawers and boxes, loaded with an eclectic combination of toys, textures, knobs, clever mechanical mechanisms, and objects that I cannot classify. I peer into the boxes, trying to figure out what they are for, what purpose they serve. "Just turn the knob," I'm told, as something is thrust into my hands. I find the knob and rotate it. It feels good: smooth, silky. I try a different knob: it doesn't feel as precise. There are dead regions where I turn and nothing seems to happen. Why the difference? Same mechanism, I am told: the difference is the addition of a special, very viscous oil. "Feel matters," a designer explains, and from the "Tech Box" appear yet more examples: silky cloth, microfiber textiles, sticky rubber, squeezable balls—more than I can assimilate at one experience.

Good designers worry a lot about the physical feel of their products. Physical touch and feel can make a huge difference in your appreciation of their creations. Consider the delights of smooth, polished metal, or soft leather, or a solid, mechanical knob that moves precisely from position to position, with no backlash or dead zones, no wobbling or wiggling. No wonder IDEO designers love their "Tech Box," their collection of toys and textures, mechanisms and controls. Many design professionals focus on visual appearance, in part because this is what can be appreciated from a distance and, of course, all that can be experienced in an advertising or marketing photograph or printed illustration. Touch and feel, however, are critical to our behavioral assessment of a product. Recall the shower of figure 3.3.

Physical objects have weight, texture, and surface. The design term for this is "tangibility." Far too many high-technology creations have moved from real physical controls and products to ones that reside on computer screens, to be operated by touching the screen or manipulating a mouse. All the pleasure of manipulating a physical object is gone and, with it, a sense of control. Physical feel matters. We are, after all, biological creatures, with physical bodies, arms, and legs. A huge amount of the brain is taken up by the sensory systems, continually probing and interacting with the environment. The best of products make full use of this interaction. Just imagine cooking, feeling the comfort of a balanced, high-quality knife, hearing the sound of cutting on the chopping board or the sizzle when you drop food into the skillet, smelling the odors released from the fresh-cut food. Or imagine gardening, feeling the tenderness of a plant, the grittiness of the earth. Or playing tennis, hearing the twang of the ball against the racket's strings, its feel in your hands. Touch, vibration, feel, smell, sound, visual appearance. And now imagine doing all this on a computer screen, where what you see may look real, but with no feel, no scent, no vibrations, no sound.

The world of software is to be commended for its power and chameleon-like ability to transform itself into whatever function is needed. The computer provides for abstract actions. Computer scientists call these environments "virtual worlds," and although they have many benefits, they eliminate one of the great delights of real interactions: the delight that comes from touching, feeling, and moving real physical objects.

The virtual worlds of software are worlds of cognition: ideas and concepts presented without physical substance. Physical objects involve the world of emotion, where you experience things, whether the comfortable sensuousness of some surfaces or the grating, uncomfortable feel of others. Although software and computers have become indispensable to daily life, too much adherence to the abstraction of the computer screen subtracts from emotional pleasure. Fortunately, some designers of many computer-based products are restoring the natural, affective pleasures of the real, tangible world. Physical controls are back in style: knobs for tuning, knobs for volume, levers for turning or switching. Hurrah!

Badly conceived behavioral design can lead to great frustration, leading to objects that have lives of their own, that refuse to obey, that provide inadequate feedback about their actions, that are unintelligible, and all in all, putting anyone who tries to use them into a big, gray funk. No wonder this frustration often erupts in rage, causes the user to kick, scream, and curse. Worse, there is no excuse for such frustration. The fault does not lie with the user; the fault lies with the design.

Why do so many designs fail? Mainly because designers and engineers are often self-centered. Engineers tend to focus upon technology, putting into a product whatever special features they themselves prefer. Many designers fail as well through their fondness for the sophisticated use of images, metaphors, and semantics that win prizes in design competitions but create products that are inaccessible to users. Web sites fail here as well, for the creators focus either upon the technical sophistication of images and sounds, or upon making sure that each division of a company receives the recognition that its political power dictates.

None of these cases takes into account the concerns of the poor user, people like you and me, who use a product or web site to satisfy some need. You need to accomplish a task or to find some information. You don't know the organizational chart of the company on whose web site you seek information, nor do you wish to. You may enjoy flashy images and sounds briefly, but not when that cleverness and sophistication get in the way of getting your job done.

Good behavioral design should be human-centered, focusing upon understanding and satisfying the needs of the people who actually use the product. As I have said, the best way to discover these needs is through observation, when the product is being used naturally, and not in response to some arbitrary request to "show us how you would do x." But observation is surprisingly rare. You would think that manufacturers would want to watch people use their products, the better to improve them for the future. But no, they are too busy designing and matching the features of the competition to find out whether their products are really effective and usable.

Engineers and designers explain that, being people themselves, they understand people, but this argument is flawed. Engineers and designers simultaneously know too much and too little. They know too much about the technology and too little about how other people live their lives and do their activities. In addition, anyone involved with a product is so close to the technical details, to the design difficulties, and to the project issues that they are unable to view the product the way an unattached person can.

Focus groups, questionnaires, and surveys are poor tools for learning about behavior, for they are divorced from actual use. Most behavior is subconscious and what people actually do can be quite different from what they think they do. We humans like to think that we know why we act as we do, but we don't, however much we like to explain our actions. The fact that both visceral and behavioral reactions are subconscious makes us unaware of our true reactions and their causes. This is why trained professionals who observe real use in real situations can often tell more about people's likes and dislikes—and the reasons for them—than the people themselves.

An interesting exception to these problems comes when designers or engineers are building something for themselves that they will use frequently in their own everyday lives. Such products tend to excel. As a result, the best products today, from a behavioral point of view, are often those that come from the athletic, sports, and craft industries, because these products do get designed, purchased, and used by people who put behavior above everything else. Go to a good hardware store and examine the hand tools used by gardeners, woodworkers, and machinists. These tools, developed over centuries of use, are carefully designed to feel good, to be balanced, to give precise feedback, and to perform well. Go to a good outfitter's shop and look at a mountain climber's tools or at the tents and backpacks used by serious hikers and campers. Or go to a professional chef's supply house and examine what real chefs buy and use in their kitchens.

I have found it interesting to compare the electronic equipment sold for consumers with the equipment sold to professionals. Although much more expensive, the professional equipment tends to be simpler and easier to use. Video recorders for the home market have numerous flashing lights, many buttons and settings, and complex menus for setting the time and programming future recording. The recorders for the professionals just have the essentials and are therefore easier to use while functioning better. This difference arises, in part, because the designers will be using the products themselves, so they know just what is important and what is not. Tools made by artisans for themselves all have this property. Designers of hiking or mountain climbing equipment may one day find their lives depending upon the quality and behavior of their own designs.

When the company Hewlett Packard was founded, their main product was test equipment for electrical engineers. "Design for the person on the next bench," was the company motto, and it served them well. Engineers found that HP products were a joy to use because they fitted the task of the electrical engineer at the design or test bench perfectly. But today, the same design philosophy no longer works: the equipment is often used by technicians and field crew who have little or no technical background. The "next bench" philosophy that worked when the designers were also users fails when the populations change.

Good behavioral design has to be a fundamental part of the design process from the very start; it cannot be adopted once the product has been completed. Behavioral design begins with understanding the user's needs, ideally derived by conducting studies of relevant behavior in homes, schools, places of work, or wherever the product will actually be used. Then the design team produces quick, rapid prototypes to test on prospective users, prototypes that take hours (not days) to build and then to test. Even simple sketches or mockups from cardboard, wood, or foam work well at this stage. As the design process continues, it incorporates the information from the tests. Soon the prototypes are more complete, sometimes fully or partially working, sometimes simply simulating working devices. By the time the product is finished, it has been thoroughly vetted through usage: final testing is necessary only to catch minor mistakes in implementation. This iterative design process is the heart of effective, user-centered design.

Reflective Design

Reflective design covers a lot of territory. It is all about message, about culture, and about the meaning of a product or its use. For one,

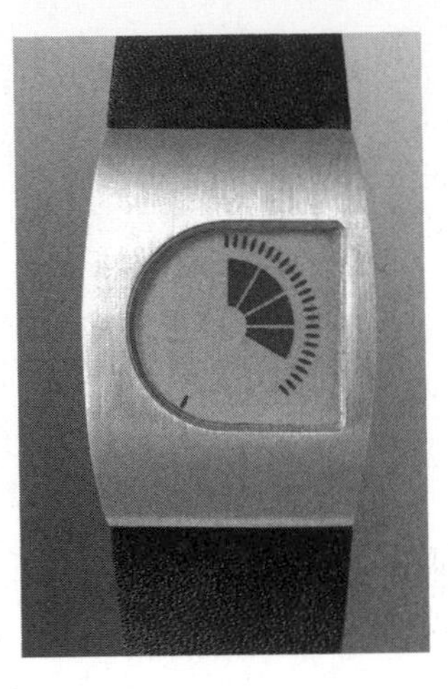

FIGURE 3.6

Reflective design through cleverness. The value of this watch comes from the clever representation of time: Quick, what time is represented? This is *Time by Design's* "Pie" watch showing the time of 4:22 and 37 seconds. The goal of the company is to invent new ways of telling time, combiing "art and time telling into amusing and thought provoking clocks and watches." This watch is as much a statement about the wearer as it is a timepiece.

(Courtesy of Time by Design.)

it is about the meaning of things, the personal remembrances something evokes. For another, very different thing, it is about self-image and the message a product sends to others. Whenever you notice that the color of someone's socks matches the rest of his or her clothes or whether those clothes are right for the occasion, you are concerned with reflective self-image.

Whether we wish to admit it or not, all of us worry about the image we present to others—or, for that matter, about the self-image that we present to ourselves. Do you sometimes avoid a purchase "because it wouldn't be right" or buy something in order to support a cause you prefer? These are reflective decisions. In fact, even people who claim a complete lack of interest in how they are perceived—dressing in whatever is easiest or most comfortable, refraining from purchasing new items until the ones they are using completely stop working—make statements about themselves and the things they care about. These are all properties of reflective processing.

Consider two watches. The first one, by "Time by Design" (figure 3.6), exhibits reflective delight in using an unusual means to display time, one that has to be explained to be understood. The watch is also viscerally attractive, but the main appeal is its unusual display. Is the time more difficult to read than on a traditional analog or digital

FIGURE 3.7

Pure behavioral design.

The *Casio* "G-Shock" watch is pure behavioral design; efficient and effective, with no claims to beauty and low in such measures of reflective design as prestige and status. But consider the behavioral aspects: two time zones, a stopwatch, a countdown timer, and an alarm. Inexpensive, easy to use, and accurate.

(Author's collection.)

watch? Yes, but it has an excellent underlying conceptual model, satisfying one of my maxims of good behavioral design: it need only be explained once; from then on, it is obvious. Is it awkward to set the watch because it has but a single control? Yes, but the reflective delight in showing off the watch and explaining its operation outweighs the difficulties. I own one myself and, as my weary friends will attest, proudly explain it to anyone who shows the slightest bit of interest. The reflective value outweighs the behavioral difficulties.

Now contrast this reflective design with the practical, sensible, plastic digital watch by Casio (figure 3.7). This is a practical watch, one emphasizing the behavioral level of design without any attributes of visceral or reflective design. This is an engineer's watch: practical, straightforward, multiple features, and low price. It isn't particularly attractive—that isn't its selling point. Moreover, the watch has no special reflective appeal, except perhaps through the reverse logic of being proud to own such a utilitarian watch when one can afford a much more expensive one. (For the record, I own both these watches, wearing the Time by Design one for formal affairs, the Casio otherwise.)

A number of years ago I visited Biel, Switzerland. I was part of a small product team for an American high-technology company, there to talk with the folks at Swatch, the watch company that had transformed the Swiss watchmaking industry. Swatch, we were proudly told, was not a watch company; it was an emotions company. Sure, they made the precision watches and movements used in most watches around the world (regardless of the brand displayed on the case), but what they had really done was to transform the purpose of a watch from timekeeping to emotion. Their expertise, their president boldly proclaimed, was human emotion, as he rolled up his sleeves to display the many watches on his arm.

Swatch is famous for transforming the watch into a fashion statement, arguing that people should own as many watches as ties, or shoes, or even shirts. You should change your watch, they proclaimed, to match mood, activity, or even the time of day. The executive team of Swatch patiently tried to explain this to us: Yes, the watch mechanism had to be inexpensive, yet of high quality and reliable (and we were very impressed by our tour of their completely automated manufacturing facilities), but the real opportunities lay in exploiting the face and body of the watch. Their web site puts it like this:

> **Swatch Is Design.** *The form of a Swatch watch is always the same.* The tiny space it offers for creative design exerts an irresistible power of attraction on artists. Why? Because the watch face and strap can take on the wildest imaginative concepts, the most unusual ideas, brilliant colors, rousing messages, art and comics, dreams for today and tomorrow, and much, much more. And that's exactly what makes each Swatch model so fascinating: it is design that incorporates a message, handwriting that bears witness to a personality.

At the time of my visit, we were impressed, but puzzled. We were technologists. The concept that a piece of advanced technology should really be thought of as a vehicle for emotions rather than for function was a bit difficult for us engineers to fathom. Our group could never get its act together enough to work in such a creative way, so nothing ever came of that venture—except for the long-lasting impression it made on me. I learned that products can be more than the sum of the functions they perform. Their real value can be in fulfilling people's emotional needs, and one of the most important needs of all is to establish one's self-image and one's place in the world. In his important book about the role of industrial design, *Watches Tell More than Time,* the designer Del Coates explains that "it is impossible, in fact, to design a watch that tells *only* time. Knowing nothing more, the design of a watch alone—or of any product—can suggest assumptions about the age, gender, and outlook of the person who wears it."

Did you ever consider buying an expensive, hand-crafted watch? Expensive jewelry? Single malt scotch or a prestige vodka? Can you really distinguish among the brands? Blind-tasting of many whiskeys, where the taster has no idea which glass contains which drink, reveals that you probably can't taste the difference. Why is an expensive original painting superior to a high-quality reproduction? Which would you prefer to have? If the painting is about aesthetics, then a good reproduction should suffice. But, obviously, paintings are more than aesthetics: they are about the reflective value of owning—or viewing—the original.

These questions are all cultural. There is nothing practical, nothing biological, about the answers. The answers are conventions, learned in whatever society you inhabit. For some of you, the answers will be obvious; for others, the questions will not even make sense. That is the essence of reflective design: it is all in the mind of the beholder.

Attractiveness is a visceral-level phenomenon—the response is entirely to the surface look of an object. Beauty comes from the reflective level. Beauty looks below the surface. Beauty comes from conscious reflection and experience. It is influenced by knowledge, learning, and culture. Objects that are unattractive on the surface can give pleasure. Discordant music, for example, can be beautiful. Ugly art can be beautiful.

Advertising can work at either the visceral or the reflective level. Pretty products—sexy automobiles, powerful-looking trucks, seductive bottles for drinks and perfume—play with the visceral level.

Prestige, perceived rarity, and exclusiveness work at the reflective level. Raise the price of Scotch, and increase the sales. Make it difficult to get reservations to a restaurant or entrance to a club, and increase their desirability. These are reflective-level ploys.

Reflective-level operations often determine a person's overall impression of a product. Here, you think back about the product, reflecting upon its total appeal and the experience of using it. Here is where many factors come into play and where the deficiencies of one aspect can be outweighed by the strengths of another. Minor difficulties might very well be overlooked in the overall assessment—or enhanced, blown all out of proportion.

The overall impact of a product comes through reflection—in retrospective memory and reassessment. Do you fondly show your possessions to friends and colleagues, or do you hide them and, if you talk at all, is it only to complain? Things that an owner is proud of will be displayed prominently, or, at the least, shown to people.

Customer relationships play a major role at the reflective level, so much so that a good relationship can completely reverse an otherwise negative experience with the product. Thus, a company that goes out of its way to assist and help disgruntled customers can often turn them into its most loyal fans. Indeed, the person who buys a product and has nothing but pleasant experiences with it may be less satisfied than the one who has an unhappy experience, but is well treated by the company as it fixes the problem. This is an expensive way to win customer loyalty, but it shows the power of the reflective level. Reflective design is really about long-term customer experience. It is about service, about providing a personal touch and a warm interaction. When a customer reflects on the product in order to decide what next to purchase or to advise friends, a pleasant reflective memory can overcome any prior negative experiences.

Amusement park rides are a good example of the interplay between reflection and reaction. The ride appeals both to those who value the feelings that accompany high arousal and fear for its own sake and to those for whom the ride is all about the reflective power afterward. At the visceral level, the whole point is to thrill riders, scaring them in the process. But this has to be done in a reassuring way. While the visceral system is operating at full force, the reflective system is a calming influence. This is a safe ride, it is telling the rest of the body. It only appears to be dangerous. It is okay. During the ride, the visceral system probably wins. But in retrospect, when memory has dimmed, the reflective system wins. Now, it is a badge of honor to have experienced the ride. It provides stories to tell other people. Here an effective amusement park enhances the interaction by selling photographs of the rider at the peak of the experience. They sell photographs and souvenirs, so the riders can brag to friends.

Would you go on a ride if the amusement park was old and shabby, with clearly broken components, rusty railings, and a general air of incompetence? Obviously not. The rational reassurance will not be nearly as effective. Once the reflective system fails, then the appeal is apt to collapse as well.

A Case Study: The National Football League Headset

"You know what the hardest part of this design was?" Walter Herbst, of the design firm Herbst LaZar Bell, asked, proudly showing me the Motorola headset (figure 3.8).

"Reliability?" I answered, hesitantly, thinking that it looked so big and strong, it must be reliable.

"Nope," he answered, "it was the coaches—making the coaches feel good about wearing it."

Motorola had asked Herbst LaZar Bell to design the headset to be used by the coaches of the National Football League. Mind you, these couldn't be just any headset. They had to be highly functional, delivering intelligible messages between coaches and their staff scattered about the stadium. The microphone boom had to be movable so that it could be placed on either side of the head for left-handed and right-

FIGURE 3.8

Motorola's headset for the coaching staff of the National Football League. The headset was designed by the industrial design firm Herbst LaZar Bell, which won a Gold Prize from both *Business Week's* Industrial Design Excellence Awards and the Industrial Design Society of America (IDSA) for their achievement. IDSA described the reasons this way:"It's a rare moment when a design team realizes that it has been given the green light to create an icon—one that will be seen by millions around the world. The Motorola NFL Headset represents the marriage of sophisticated communications technology and great design with the blood, sweat and tears on the field of play. In addition, it enhances awareness of a company committed to delivering on the demanding requirements of professional users in every arena." *(Courtesy of Herbst LaZar Bell and Motorola, Inc.)*

handed coaches. The environment is a difficult one. It is noisy. Football games are played in temperature extremes, from high heat, to rain, to extreme cold. And headsets get abused: angry coaches take out their frustration on whatever happens to be around, sometimes grabbing hold of the microphone boom and throwing the headset to the ground. The signals have to be private so opposing teams cannot eavesdrop. The headset is also an important advertising symbol, exposing the Motorola name to television viewers, so the brand name has to be visible regardless of camera angle. And finally, the coaches have to be satisfied. They have to want to use it. So not only does the headset have to stand up to the rigors of the game, it has to be comfortable to wear for hours at a time.

The headset design was a challenge. Small, lightweight headsets, though more comfortable, are not strong enough. More importantly, though, coaches rejected them. The coaches are the leaders of a large, active team. Football players are among the largest, most muscular players in team sports. The headset had to reinforce this image: it had to be muscular itself to convey the image of a coach in charge of things.

So, yes, the design had to have visceral appeal; and, yes, it had to meet the behavioral objectives. The biggest challenge, however, was to do all this while satisfying the coaches, projecting the heroic, manly self-image of strong, disciplined leaders who managed the world's toughest players, and who were always in control. In short: reflective design.

Accomplishing all this took a lot of work. This was not a design to be scribbled on the side of a napkin (although a lot of trial designs were, in fact, done on napkins). Sophisticated computer-aided drawing tools that allowed the designers to visualize just how the headset looked from all angles before anything had been built, optimizing the interaction of ear cups and microphone, headband adjustment, and even the placement of the logos (maximizing their visibility to the TV audience while simultaneously minimizing it to the coaches, to avoid distraction).

"The main goal in designing the Coaches Headset," said Steve Remy, project manager for Herbst LaZar Bell, "was to create a cool new look for the product that is often overlooked as a background item, and turn it into an image-building product that attracts the viewer's attention even in the high energy, action-packed context of the professional football game." It worked. The result is a "cool" product—one that not only functions well, but also serves as an effective advertising tool for Motorola and enhances the self-image of the coaches. This is an excellent example of how the three different aspects of design can work well with one another.

The Devious Side of Design

> To the uninitiated, walking into the Diesel jeans store on Union Square West feels a lot like stumbling into a rave. Techno music pounds at a mind-rattling level. A television plays a videotape of a Japanese boxing match, inexplicably. There are no helpful signs pointing to men's or women's departments, and no obvious staff members in sight.
>
> While large clothing retailers like Banana Republic and Gap have standardized and simplified the layout of their stores in an effort to put customers at ease, Diesel's approach is based on the unconventional premise that the best customer is a disoriented one.
>
> "We're conscious of the fact that, outwardly, we have an intimidating environment," said Niall Maher, Diesel's director of retail operations. "We didn't design our stores to be user-friendly because we want you to interact with our people. You can't understand Diesel without talking to someone."
>
> Indeed, it is at just the moment when a potential Diesel customer reaches a kind of shopping vertigo that members of the company's intimidatingly with-it staff make their move. Acting as salesmen-in-shining-armor, they rescue—or prey upon, depending on one's point of view—wayward shoppers.
>
> —Warren St. John, *New York Times*

To the practitioner of human-centered design, serving customers means relieving them of frustration, of confusion, of a sense of helplessness. Make them feel in control and empowered. To the clever salesperson, just the reverse is true. If people don't really know what they want, then what is the best way to satisfy their needs? In the case of human-centered design, it is to provide them with the tools to explore by themselves, to try this and that, to empower themselves to success. To the sales staff, this is an opportunity to present themselves as rescuers "in-shining-armor," ready to offer assistance, to provide just the answer customers will be led to believe they had been seeking.

In the world of fashion—which encompasses everything from clothes to restaurants, automobiles to furniture—who is to say which approach is right, which wrong? The solution through confusion is a pure play on emotions, selling you, the customer, the idea that the proposed item will precisely serve your needs and, more important, advertise to the rest of the world what a superior, tasteful, "with it" person you are. And, if you believe it, it will probably come to pass, for strong emotional attachment provides the mechanism for self-fulfilling prophecy.

So, again, which approach is right: that of the Gap and Banana Republic, which "have standardized and simplified the layout of their stores in an effort to put customers at ease"; or Diesel, which deliberately confuses and intimidates, the better to prepare the customer to welcome the helpful, reassuring salesperson? I know my preferences; I'll go with Gap and Banana Republic any day, but the very success of Diesel shows that not everyone shares this view. In the end, the stores serve different needs. The first two stores are more utilitarian (although they would shudder to be called that); the second pure fashion, where the whole goal is caring about what others think.

"When you're wearing a thousand-dollar suit," super salesman Mort Spivas told the media critic Douglass Rushkoff, "you project a different aura. And then people treat you differently. You exude confidence. And if you can feel confident, you'll *act* confident." If salespeople believe that wearing an expensive suit makes them different, then it does make them different. For fashion, emotions are key. Stores that manipulate emotions are simply playing the game consumers have invited themselves into. Now, the fashion world may have inappropriately brainwashed the eager public into believing that the game counts, but that is the belief, nonetheless.

To disconcert shoppers as a selling tool is hardly news. Supermarkets long ago learned to put the most frequently desired items at the rear of the store, forcing buyers to pass by isles of tempting impulse purchases. Moreover, related items can be placed nearby. Do people rush to the store to buy milk? Put the milk at the rear of the store, and put cookies nearby. Do they rush in to buy beer? Put the beer next to snacks. Similarly, at the checkout counter, display the small, last-minute items people might be tempted to buy while waiting in line. Creating these "point of purchase" displays has become a big business. I can even imagine stores deliberately slowing up the checkout procedure to give customers more time to make those last-minute, impulse purchases.

Once a customer has learned the shop or shelf layout, it is time to redo it, goes this marketing philosophy. Otherwise, a shopper wanting a can of soup will simply go directly to the soup and not notice any of the other enticing items. Rearranging the store forces the shopper to explore previously unvisited aisles. Similarly, rearranging how the soups are stored prevents the shopper from buying the same type of soup each time without ever trying any other variety. So shelves get rearranged, and related items are put nearby. Stores get restructured, and the most popular items are placed at the furthest ends of the store, with impulse items placed either adjacent or at the "end caps," the ends of the aisles where they are most visible. There is a perverse set of usability principles at play here: make it difficult to buy the most desired items, and extremely easy for the impulse items.

When these tricks are used, it is critically important that the shopper not notice. Make the store layout appear normal. Indeed, make the disorientation part of the fun. Diesel gets away with their confusion because they are famous for it, because their clothes are very popular, and because wandering through the store is part of the experience. The same philosophy would not work for a hardware store. In the supermarket, the fact that milk or beer is at the farthest end of the store doesn't appear deliberate, it seems natural. After all, the coolers are there at the back, which is where these items are kept. Of course, no one ever asks the real question, Why are the coolers located there?

Once shoppers realize that they are being manipulated in these ways, a backlash may occur whereby shoppers desert the manipulative stores and visit the ones that make their experiences more pleasant. Stores that try to profit through confusion often enjoy a meteoric rise in sales and popularity, but suffer a similar meteoric fall as well. The staid, conventional, helpful store is more stable, with neither the great ups nor the great downs in popularity. Yes, shopping can be a sensual, emotional experience, but it can also be a negative, traumatic one. But when stores do things correctly, when they understand "The Science of Shopping," to use the subtitle of Paco Underhill's book, then the experience can be both a positive emotional one for the shopper and a profitable one for the seller.

Just as the scary rides of an amusement park pit the anxiety and fear of the visceral level against the calm reassurance of the intellect, the Diesel store pits the initial confusion and anxiety at both the behavioral and reflective levels against the relief and welcome of the rescuing sales person. In both cases, the initial negative affect is necessary to set up the relief and delight at the end. In the park, the ride is now safely over, and the rider can reflect upon all the positive experiences of having successfully mastered the adventure. In the store, the relieved customer reflects back upon the calm guidance and reassurance offered by the salesperson. In the store, the customer is apt to bond with the salesperson, not unlike the "Stockholm syndrome," in which kidnap victims develop such a positive emotional bond with their captors that, after they are freed and the captors in custody, they plead for mercy for the kidnappers. (The name comes from a bank robbery in the early 1970s in Stockholm, Sweden, where a hostage developed a romantic attachment to one of her captors.) But there is a real difference between these two cases. In the amusement park, the fear and excitement is the draw. It is public, advertised. In the Diesel store, it is artificial, manipulative. One is natural, the other not. Guess which will last over time.

Design by Committee Versus by an Individual

Although reflective thought is the essence of great literature and art, films and music, web sites and products, appealing to the intellect is no guarantee of success. Many well-acclaimed serious works of art and music are relatively unintelligible to the average person. I suspect they may even be unintelligible to those who proclaim them, for in the exalted realm of literature, art, and professional criticism, it would appear that when something can be clearly understood, it is judged as flawed, whereas when something is impenetrable, it must of necessity be good. And some items convey such subtle, hidden intellectual messages, that they are lost on the average viewer or user, perhaps lost on everyone except the creator and dutiful students in universities, listening to the learned critiques from their professors.

Consider the fate of Fritz Lang's classic film "Metropolis," "a wildly ambitious, hugely expensive science fiction allegory of filial revolt, romantic love, alienated labor and dehumanizing technology." It was first shown in Berlin in 1926 but the American distributor, Paramount Films, complained that it was unintelligible. They hired Channing Pollock, a playwright, to reedit the film. Pollock complained that "symbolism ran such riot that people who saw it couldn't tell what the picture was all about." Whether or not one agrees with Pollock's criticism, there is no doubt that too much intellectualism can certainly get in the way of pleasure and enjoyment. (Which, of course, is often beside the point: The purpose of a serious essay, movie, or piece of art is to educate and inform, not to amuse.)

There is a fundamental conflict between the preferences of the popular audience and the desires of the intellectual and artistic community. The case is most easily made with respect to movies, but also applies to all design as well as to serious music, art, literature, drama, and television.

Making a movie is a complex process. Hundreds of people are ultimately involved, with layers of producers, directors, screenwriters, cameramen, editors, and studio executives all having some legitimate say in the end product. Artistic integrity, a cohesive thematic approach, and deep substance seldom come from committees. The best designs come from following a cohesive theme throughout, with a clear vision and focus. Usually, such designs are driven by the vision of one person.

You may think that I am contradicting one of my standard design rules: test and redesign, test and redesign. I have long championed human-centered design, where a product undergoes continual revision based upon tests with potential users of the product. This is a time-tested, effective method for producing usable products whose end result fits the needs of the largest number of people. Why do I now claim that a single designer who has a clear model of the end product and ensures that it gets developed can be superior to that cautious design cycle of design, test, and then redesign?

The difference is that all my previous work focused upon behavioral design. I still maintain that an iterative, human-centered approach works well for behavioral design, but it is not necessarily appropriate for either the visceral or the reflective side. When it comes to these levels, the iterative method is design by compromise, by committee, and by consensus. This guarantees a result that is safe and effective, but invariably dull.

This is what happens with movies. Movie studio executives often subject movies to screen tests, where a film is shown to a test audience and their reactions gauged. As a result, scenes are deleted, story lines changed. Frequently an ending is changed to make it more comfortable to the viewers. All of this is done to increase the popularity and sales of the movie. The problem is that the director, cameramen, and writers are apt to feel that the changes have destroyed the soul of the film. Who to believe? I suspect both the test results and the opinions of the creative crew are valid.

Films are judged by a variety of standards. On the one hand, even an "inexpensive" film can cost millions of dollars to produce, while an expensive one can cost hundreds of millions. A film can be both a major business investment and an artistic statement.

Business versus art or literature: the debate is real and appropriate. In the end, the decision is whether one wishes to be an artist making a statement, in which case profits are irrelevant, or a business person, changing the film or product to make it appeal to as many people as possible, even at the cost of artistic merit. Want a popular film, one that appeals to the masses? Show the film to test audiences and revise. Want an artistic masterpiece? Hire a great creative crew that you can trust.

Henry Lieberman, a research scientist at the MIT Media Laboratory has described the case against "design by committee" most eloquently, so let me simply repeat his words here:

> The brilliant conceptual artists Vitaly Komar and Alex Melamid conducted surveys asking people questions like, What's your favorite color? Do you prefer landscapes to portraits? Then they produced exhibitions of perfectly "user-centered art." The results were profoundly disturbing. The works were completely lacking in innovation or finesse of craftsmanship, disliked even by the very same survey respondents. Good art is not an optimal point in a multidimensional space; that was, of course, their point. Perfectly "user-centered design" would be disturbing as well, precisely because it would lack that artistry.

One thing is certain, this debate is fundamental: it will continue as long as the creators of art, music, and performance are not the same people as those who must pay to get them distributed to the world. If you want a successful product, test and revise. If you want a great product, one that can change the world, let it be driven by someone with a clear vision. The latter presents more financial risk, but it is the only path to greatness.

CHAPTER FOUR

Fun and Games

PROFESSOR HIROSHI ISHII of the MIT Media Laboratory runs back and forth, eager to show me all his exhibits. "Pick up a bottle," he says, standing in front of a colorfully lit stand of glass bottles. I do so and am rewarded by a playful tune. I pick up a second bottle and another instrument joins in, playing in harmony with the first. Pick up the third bottle, and the instrumental trio is complete. Put down one of the bottles and the instrument associated with it stops. I'm intrigued, but Hiroshi is anxious for me to experience more. "Here, look at this," Hiroshi is calling from the other side of the room, "try this." What is going on? I don't know, but it certainly is fun. I could spend the whole day there.

But Hiroshi has more delights to show. Imagine trying to play table tennis on a school of fish, as in figure 4.1. There they are, swimming about the table, their images delivered by a projector located in the ceiling above the table. Each time the ball hits the table surface, the ripples spread out and the fish scatter. But the fish can't get away—it's a small table, and no matter where the fish go, the ball soon scatters

FIGURE 4.1

Table tennis on top of a school of fish.

"Ping Pong Plus." Images of water and a school of fish are projected onto the surface of the ping pong table. Each time the ball hits the table, the computer senses its position, causing the images of ripples to spread out from the ball and the fish to scatter.

(Courtesy of Hiroshi Ishii of the MIT Media Laboratory.)

them again. Is this a good way to play table tennis? No, but that's not what it's about: it's about fun, delight, the pleasure of the experience.

Fun and pleasure, alas, are not topics often covered by science. Science can be too serious, and even when it attempts to examine the issues surrounding fun and pleasure, its very seriousness becomes a distraction. Yes, there are conferences on the scientific basis of humor, of fun ("funology" is the name given to this particular endeavor), but this is a difficult topic and progress is slow. Fun is still an art form, best left to the creative minds of writers, directors, and other artists. But the lack of scientific understanding should not get in the way of our enjoyment. Artists often pave the way, exploring approaches to human interaction that science then struggles to understand. This has long been true in drama, literature, art, and music, and it is these areas that provide lessons for design. Fun and games: a worthwhile pursuit.

Designing Objects for Fun and Pleasure

Why must information be presented in a dull, dreary fashion, such as in a table of numbers? Most of the time we don't need actual numbers, just some indication of whether the trend is up or down, fast or slow, or some rough estimate of the value. So why not display the information in a colorful manner, continually available in the periphery of attention, but in a way that delights rather than distracts? Once again, Professor Ishii suggests the means: Imagine colorful pinwheels spinning above your head, enjoyable to contemplate, but where the rate of spin is meaningful, perhaps coupled to the outside temperature, or maybe volume of traffic on the roads you use for your daily commute, or for any statistic that is useful to watch. Do you need to be reminded to do something at a specific time? Why not have the pinwheels increase their speed as the time approaches, the higher rate of speed being more likely to attract your attention and, simultaneously, to indicate the urgency. Spinning pinwheels? Why not? Why not have information displayed in a pleasant, comfortable way?

Technology should bring more to our lives than the improved performance of tasks: it should add richness and enjoyment. A good way to bring fun and enjoyment to our lives is to trust in the skill of artists. Fortunately, there are many around.

Consider the pleasure of the Japanese lunchbox, which started as a simple work lunch. In the box lunch you can enjoy a wide assortment of foods, wide enough so that even if you do not like some of the entrées, there are other choices. The box is small, yet fully packed, which poses an aesthetic challenge to the chef. In the best of cases

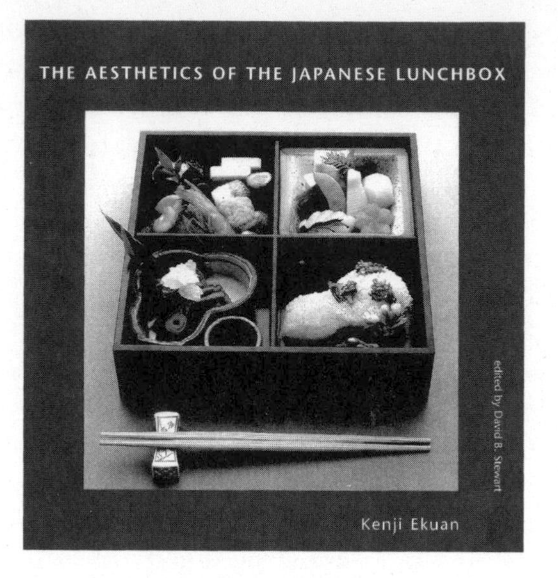

FIGURE 4.2

The cover of Kenji Ekuan's book

The Aesthetics of the Japanese Lunchbox.

The book illustrates how design should incorporate depth, beauty, and utility. Ekuan demonstrates that the lunchbox is a metaphor for much of Japanese design philosophy. It is art meant to be consumed. It follows the philosophy more is better, offering an assortment of foods so that everyone can find something to their taste. It originated as a practical, working person's lunch, so it combines function, practicality, and beauty—as well as an exercise in philosophy.

(Photograph by Takeshi Doi, with permission of Doi, Ekuan and MIT Press.)

(figure 4.2), the result is a work of art: art meant to be consumed. The Japanese industrial designer Kanji Ekuan has suggested that the aesthetics of the Japanese lunchbox is an excellent metaphor for design. This lunchbox, divided into small compartments, each with five or six types of food, packs twenty to twenty-five colors and flavors within its small space. Ekuan describes it this way:

> the cook . . . would naturally be disappointed if the result of such an effort were eaten without a glance or a second thought, (so) he or she

> works to make lunchbox meals so attractive that guests are actually reluctant to take up their chopsticks and begin eating. But even so, it is only a matter of time before the masterwork is consumed. The guest senses the formal layout even as he proceeds to break up the perfected layout. This is the inherent and paradoxical relationship between the provision and the acceptance of beauty.

The crowded nature of the lunchbox has many virtues. It forces attention to detail in the arrangement and presentation of the food. This essence of design, packing a lot into a small space while maintaining an aesthetic sense, says Ekuan, is the essence of much of Japan's design for high technology, where one goal is "to establish multifunctionality and miniaturization as equal values. Packing numerous functions into something and making it smaller and thinner are contradictory aims, but one had to pursue contradiction to the limit to find a solution."

The trick is to compress multiple functions into limited space in a way that does not compromise the various dimensions of design. Ekuan clearly prizes beauty—aesthetics—first. "A sense of beauty that lauds lightness and simplicity," he continues, "desire that precipitates functionality, comfort, luxury, diversity. Fulfillment of beauty and its concomitant desire will be the aim of design in the future."

Beauty, fun, and pleasure all work together to produce enjoyment, a state of positive affect. Most scientific studies of emotion have focused upon the negative side, upon anxiety, fear, and anger, even though fun, joy, and pleasure are the desired attributes of life. The climate is changing, with articles and books on "positive psychology" and "well-being" becoming popular. Positive emotions trigger many benefits: They facilitate coping with stress. They are essential to people's curiosity and ability to learn. Here is how the psychologists Barbara Fredrickson and Thomas Joiner describe positive emotions:

> positive emotions *broaden* people's thought-action repertoires, encouraging them to discover novel lines of thought or action. Joy, for

instance, creates the urge to play, interest creates the urge to explore, and so on. Play, for instance, builds physical, socioemotional, and intellectual skills, and fuels brain development. Similarly, exploration increases knowledge and psychological complexity.

It doesn't take much to transform otherwise dull data into a bit of fun. Contrast the style of three major internet search companies. Google stretches out its logo to fit the number of results in a playful, jolly way (figure 4.3). Several people have told me how much they look forward to seeing just how long the *Gooooogle* will get. But Yahoo, Microsoft network (MSN), and many other sites forgo any notion of fun, and instead present the straightforward results in an unimaginative, orderly way. Small point? Yes, but a meaningful one. Google is known as a playful, fun site—as well as a very useful one—and this playful distortion of its logo helps reinforce this brand image: Fun for the user of the site, good reflective design, and good for business.

The academic, research enterprise of design has not done a good job of studying fun and pleasure. Design is usually thought of as a practical skill, a profession rather than a discipline. In my research for this book, I found lots of literature on behavioral design, much discussion of aesthetics, image, and advertising. The book *Emotional*

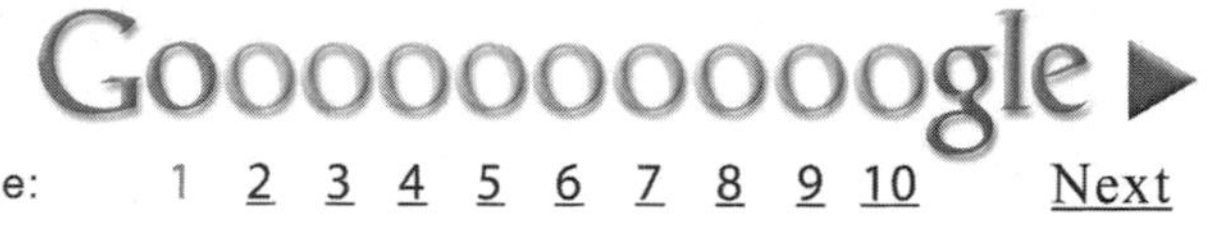

FIGURE 4.3

Google plays with their name and logo in a creative, inspiring way. Some searches return multiple pages, so Google modifies its logo accordingly: When I performed a search on the phrase "emotion and design" I got 10 pages of results. Google stretched its logo to put 10 "Os" in its name, providing some fun while also being informative and, best of all, non-intrusive. *(Courtesy of Google.)*

Branding is a treatment of advertising, for example. Academics have concentrated primarily upon the history of design, or the social history or societal implications, or if they are from the cognitive and computer sciences, upon the study of machine interfaces and usability.

In *Designing Pleasurable Products*, one of the few scientific studies of pleasure and design, the human factors expert and designer Patrick Jordan builds on the work of Lionel Tiger to identify four kinds of pleasure. Here is my interpretation:

Physio-pleasure. Pleasures of the body. Sights, sounds, smells, taste, and touch. Physio-pleasure combines many aspects of the visceral level with some of the behavioral level.

Socio-pleasure. Social pleasure derived from interaction with others. Jordan points out that many products play an important social role, either by design or by accident. All communication technologies—whether telephone, cell phone, email, instant messaging, or even regular mail—play important social roles by design. Sometimes the social pleasure derives serendipitously as a byproduct of usage. Thus, the office coffeemaker and mailroom serve as focal points for impromptu gatherings at the office. Similarly, the kitchen is the focal point for many social interactions in the home. Socio-pleasure, therefore, combines aspects of both behavioral and reflective design.

Psycho-pleasure. This aspect of pleasure deals with people's reactions and psychological state during the use of products. Psycho-pleasure resides at the behavioral level.

Ideo-pleasure. Here lies the reflection on the experience. This is where one appreciates the aesthetics, or the quality, or perhaps the extent to which a product enhances life and respects the environment. As Jordan points out, the value of many products comes from the statement they make. When displayed so that others can see them, they provide ideo-pleasure to the extent that they signify the value judgments of their owner. Ideo-pleasure clearly lies at the reflective level.

Take the Jordan/Tiger classification, mix with equal parts of the three design levels, and you have a fun and pleasurable end result. But fun and pleasure are elusive concepts. Thus, what is considered delightful depends a lot upon the context. The actions of a kitten or human baby may be judged fun and cute, but the very same actions performed by a cat or human adult can be judged irritating or disgusting. Moreover, what is fun at first can outwear its welcome.

Consider the "Te ò" tea strainer (figure 4.4), designed by Stefano Pirovano for the Italian manufacturing firm Alessi. At first glance it is cute, childish even. As such, it doesn't qualify as fun—not yet. It is a simple animate figure. The day I purchased it, I had lunch with Keiichi Sato, a professor of design at the Illinois Institute of Technology's Institute of Design in Chicago. At the lunch table, I proudly displayed my new purchase. Sato's first response was skeptical. "Yes," he said, "it's pleasant and cute, but to what purpose?" But when I placed the strainer on a cup, his eyes lit up, and he laughed (see figure 4.5).

At first sight, the arms and legs of the figure are simply cute, but when it becomes apparent that the cuteness is also functional, then "cute" becomes transformed into "pleasure" and "fun," and this, moreover, is long-lasting. Sato and I spent much of the next hour trying to understand what transforms an impression of shallow cuteness into one of deep, long-lasting pleasure. In the case of the *Te ò* strainer, the unexpected transformation is the key. Both of us noted that the essence of the surprise was the separation between the two viewings: first the tea strainer alone, then on the teacup. "If you publish this in your book," Sato warned me, "make sure that you have only the picture of the strainer visible on one page, and make the reader turn the page to see the strainer on a teacup. If you don't do that, the surprise—and the fun—will not be as strong." As you see, I have followed that advice.

What transforms the strainer from "cute" into "fun"? Is it surprise? Cleverness? Certainly both of those traits play a big role.

Does familiarity breed contempt, as the old folk saying would have

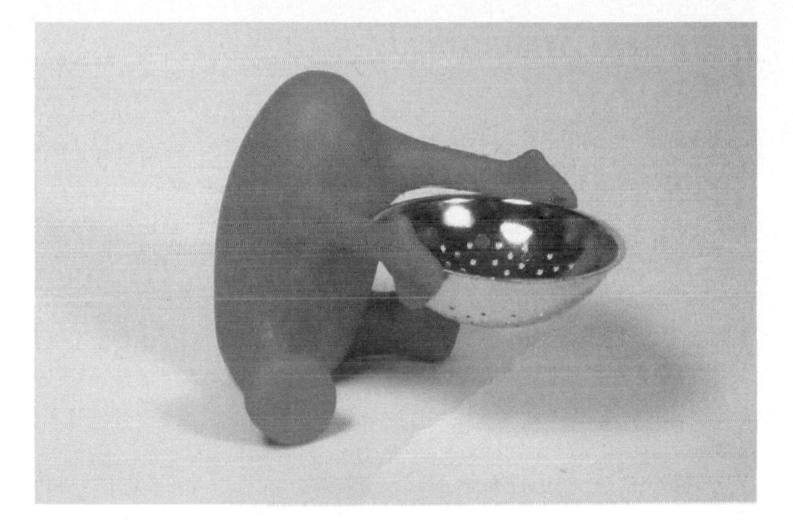

FIGURE 4.4

Stefano Pirovano's Te ò tea strainer, made by Alessi.

The figure is cute, the color and shapes attractive. Pleasureable? Yes, slightly. Fun? Not yet.

(Author's collection.)

it? Many things are cute or fun at first, but over time diminish or even become tiresome. In my home the tea strainer is now permanently on display, perched on a teacup nestled among the three teapots on the window shelf of my kitchen. The charm of the tea strainer is that it retains its fun even after considerable use, even though I see it everyday.

Now, the tea strainer is just a trifle, and I do not believe that even Pirovano, its designer, would disagree. But it passes the test of time. This is one of the hallmarks of good design. Great design—like great literature, music, or art—can be appreciated even after continual use, continued presence.

People tend to pay less attention to familiar things, whether it's a possession or even a spouse. On the whole, this adaptive behavior is biologically useful (for objects, events, and situations, not for spouses), because it is usually the novel, unexpected things in life that

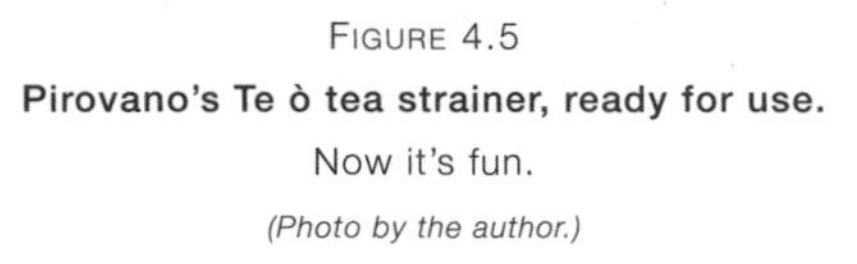

FIGURE 4.5
Pirovano's Te ò tea strainer, ready for use.
Now it's fun.
(Photo by the author.)

require the most attention. The brain naturally adapts to repeated experiences. If I were to show you a series of repeated images and measure your brain responses, the activity would diminish with the repetitions. Your brain would respond again only when something new was presented. Scientists have shown that the biggest responses always come with the least expected event. A simple sentence such as, "He picked up the hammer and nail" gives a tiny response; change the last few words, "He picked up the hammer and ate it," and you'll see a much larger one.

Human adaptation creates a challenge for design, but an opportunity for manufacturers: When people tire of an item, perhaps they will buy a new one. Indeed, the essence of fashion is to make the current trends obsolete and boring, turning them into yesterday's favorites. The attractive appliance of yesterday no longer looks quite as attractive today. Some of the examples in this book may have already followed this trajectory: the Mini Cooper automobile, so charming and cute to the reviewers at the time of this book's writing, may look dated, old-fashioned, and dull by the time you flip through these pages—so much so that you may wonder how I came to choose it as an example.

The concern for the diminishing impact of familiarity has led some designers to propose hiding beautiful views, lest continual encounter might diminish their emotional impact. In the book *A Pattern Language*, the architect Christopher Alexander and his colleagues describe 253 different design patterns derived from their observations and analyses. These patterns provide the basis of their guidelines for "a timeless way of building," which structures buildings in ways calculated to enhance the experience of the people living within them. Pattern number 134 deals with the problem of overexposure:

> *Pattern 134: Zen View*. If there is a beautiful view, don't spoil it by building huge windows that gape incessantly at it. Instead, put the windows which look onto the view at places of transition—along paths, in hallways, in entry ways, on stairs, between rooms.

If the view window is correctly placed, people will see a glimpse of the distant view as they come up to the window or pass it: but the view is never visible from the places where people stay.

The name "Zen View" comes from "the parable of a Buddhist monk who lived on a mountain with a beautiful view. The monk built a wall that obscured the view from every angle, except for a single fleeting glimpse along the walk up to his hut." In this way, said Alexander and colleagues, "the view of the distant sea is so restrained that it stays alive forever. Who, that has ever seen that view, can ever forget it? Its power will never fade. Even for the man who lives there, coming past that view day after day, it will still be alive."

Most people, however, are not Buddhist monks. Most of us would be unable to resist the temptation to engulf ourselves in such beauty. Whether hiding beauty is appropriate for all of us is up for debate, and although the fable described as the rationale for the Zen view is interesting, it is opinion, not fact. Given the chance to experience beauty for some period of time, is the total enhancement greater if the beauty is always there to be appreciated, even if it does fade with time? Or is the enhancement greater when it can only be glimpsed now and then? I do not think anyone knows the answer to this query.

I, for one, go straight for immediate enjoyment. I have always built my homes with large windows facing a view (ocean, when I lived in Southern California; pond with geese, ducks, and herons, when I lived in Northern Illinois), so I am not ready to endorse pattern 134, the Zen View, as a universal design principle.

The issue, however, is a real one. How can we maintain excitement, interest, and aesthetic pleasure for a lifetime? I suspect that part of the answer will come from the study of those things that do stand the test of time, such as some music, literature, and art. In all these cases, the works are rich and deep, so that there is something different to be perceived on each experience. Consider classical music. For many it is boring and uninteresting, but for others it can indeed be listened to with enjoyment over a lifetime. I believe that this longevity derives from the richness and complexity of its structure. The music interleaves multiple themes and variations, some simultaneous, some sequential. Human conscious attention is limited by what it can attend to at any moment, which means that consciousness is restricted to a limited subset of the musical relationships. As a result, each new listening focuses upon a different aspect of the music. The music is never boring because it is never the same. I believe a similar analysis will reveal similar richness for all experiences that last: classical music, art, and literature. So, too, with views.

The views I treasure are dynamic. Scenes are continually changing. The vegetation changes with the seasons, the lighting with the time of day. Different animals congregate at different times, and their interactions with one another and with the environment are ever-changing. In California, the waves rolling in from the ocean change continually, reflecting weather patterns from thousands of miles away. The various sea animals visible from my windows—brown pelicans, gray whales, the black-suited surfers, and dolphins—varied their activities according to the weather, the time, and the activities of those around them. Why wasn't the Zen view just as rich, just as long lasting?

Maybe the problem lies not in the object being viewed but in the viewer. It's quite possible the Buddhist monk had never learned to look. For once you have learned how to look at, listen to, and analyze what is before you, you realize that the experience is ever changing. The pleasure is forever.

This conclusion has two important implications. First, the object must be rich and complex, one that gives rise to a never-ending interplay among the elements. Second, the viewer must be able to take the time to study, analyze, and consider such rich interplay; otherwise, the scene becomes commonplace. If something is to give lifelong pleasure, two components are required: the skill of the designer in providing a powerful, rich experience, and the skill of the perceiver.

How may a design maintain its effectiveness even after long acquaintance? The secret, say designers Julie Khaslavsky and Nathan Shedroff, is seduction.

> The seductive power of the design of certain material and virtual objects can transcend issues of price and performance for buyers and users alike. To many an engineer's dismay, the appearance of a product can sometimes make or break the product's market reaction. What they have in common is the ability to create an emotional bond with their audiences, almost a need for them.

Seduction, Khaslavsky and Shedroff argue, is a process. It gives rise to a rich and compelling experience that lasts over time. Yes, there has to be an initial attraction. But the real trick—and where most products fail—is in maintaining the relationship after that initial burst of enthusiasm. Make something consciously cute where the cuteness is extraneous, irrelevant to the task, and you get frustration, irritation, and resentment. Think how many gadgets or items of furniture you

FIGURE 4.6

Two items of seduction.

Philippe Starck's "Juicy Salif" citrus juicer alongside my Global kitchen knife. Rotate the orange half on the ribbed top of the juicer and the juice flows down the sides and drips from the point into the glass. Except this gold-plated version will be damaged by the acidic fluid. As Starck is rumored to have said, *"My juicer is not meant to squeeze lemons; it is meant to start conversations."* (Author's collection.)

have brought home excitedly and then, after the first use or two, banished to storage. How many survive the passage of time and are still in use, still giving joy? And what is the difference between these two experiences?

Khaslavsky and Shedroff suggest that the three basic steps are *enticement*, *relationship*, and *fulfillment*: make an emotional promise, continually fulfill the promise, and end the experience in a memorable way. They illustrate their argument by examining the citrus juicer designed by Philippe Starck (figure 4.6). The juicer, whose full name is "Juicy Salif," was designed on a napkin in a pizza parlor in Capraia, an island in Tuscany, Italy. Alberto Alessi, whose company manufacturers them, describes the design this way:

> On the napkin, along with some incomprehensible marks (tomato sauce, in all likelihood) there were some sketches. Sketches of squids. They started on the left and, as they worked their way over to the right, they took on the unmistakable shape of what was to become the most celebrated citrus-fruit squeezer of the century that has just come to a close. You can imagine what happened: while eating a dish of squid and squeezing a lemon over it, our man had finally received his inspiration! Juicy Salif was born, and with it some headaches for the champions of "Form follows function."

That juicer was indeed seductive. I saw it and immediately went through the sequence of responses so loved by merchants: "Wow, I want it," I said to myself. Only then did I ask, "What is it? What does it do? How much does it cost?" concluding with "I'll buy it," which I did. That was pure visceral reaction. The juicer is indeed bizarre, but delightful. Why? Fortunately, Khaslavsky and Shedroff have done the analysis for me:

> **Entices by diverting attention.** It is unlike every other kitchen product by nature of its shape, form, and materials.
>
> **Delivers surprising novelty.** It is not immediately identifiable as a juicer, and its form is unusual enough to be intriguing, even surprising when its purpose first becomes clear.
>
> **Goes beyond obvious needs and expectations.** To satisfy these criteria—of being surprising and novel—it need only be bright orange or all wood. It goes so far beyond what is expected or required, it becomes something else entirely.
>
> **Creates an instinctive response.** At first, the shape creates curiosity, then the emotional response of confusion and, perhaps, fear, since it is so sharp and dangerous looking.

Espouses values or connections to personal goals. It transforms the routine act of juicing an orange into a special experience. Its innovative approach, simplicity, and elegance in shape and performance creates an appreciation and the desire to possess not only the object but the values that helped create it, including innovation, originality, elegance, and sophistication. It speaks as much about the person who owns it as it does about its designer.

Promises to fulfill these goals. It promises to make an ordinary action extraordinary. It also promises to raise the status of the owner to a higher level of sophistication for recognizing its qualities.

Leads the casual viewer to discover something deeper about the juicing experience. While the juicer doesn't necessarily teach the user anything new about juice or juicing, it does teach the lesson that even ordinary things in life can be interesting and that design can enhance living. It also teaches to expect wonder where it is unexpected—all positive feelings about the future.

Fulfills these promises. Every time it is used, it reminds the user of its elegance and approach to design. It fulfills these promises through its performance, reconjuring the emotions originally connected with the product. It also serves as a point of surprise and conversation by the associates of its owner—and is another chance to espouse its values and have them validated.

However compelling this analysis of the juicer as an item of seduction, it leaves out one important component: the reflective joy of explanation. The juicer tells a story. Anyone who owns it has to show it off, to explain it, perhaps to demonstrate it. But mind you, the juicer is not really meant to be used to make juice. As Starck is rumored to have said, "My juicer is not meant to squeeze lemons; it is meant to start conversations." Indeed, the version I own, the expensive, numbered, special anniversary edition (gold plated, no less), is explicit: "It is not intended to be used as a juice squeezer," says the numbered card attached to the juicer. "The gold plating could be damaged if it comes into contact with anything acidic."

I bought an expensive juicer, but I am not permitted to use it for making juice! Score zero for behavioral design. But so what? I proudly display the juicer in my entrance hall. Score one hundred for visceral appeal. Score one hundred for reflective appeal. (But I did use it once for making juice—who could resist?)

Seduction is real. Take the Global kitchen knife of figure 4.6, shown alongside the juicer. Unlike the juicer, which is primarily an object for display, not for use, the knife is beautiful to look at and a joy to use. It is well balanced, it feels good to the hand, and it is sharper than any other knife I have ever owned. Seduction indeed! I look forward to cutting when I cook, for these knives (I own three different types) fulfill all the requirements of seduction put forth by Khaslavsky and Shedroff.

Music and Other Sounds

Music plays a special role in our emotional lives. The responses to rhythm and rhyme, melody and tune are so basic, so constant across all societies and cultures that they must be part of our evolutionary heritage, with many of the responses pre-wired at the visceral level. Rhythm follows the natural beats of the body, with fast rhythms suitable for tapping or marching, slower rhythms for walking, or swaying. Dance, too, is universal. Slow tempos and minor keys are sad. Fast, melodic music that is danceable, with harmonious sounds and relatively constant ranges of pitch and loudness, is happy. Fear is expressed with rapid tempos, dissonance, and abrupt changes in loudness and pitch. The whole brain is involved—perception, action, cognition, and emotion: visceral, behavioral, and reflective. Some aspects of music are common to all people; some vary greatly from culture to culture. Although the neuroscience and psychology of music are widely studied, they are still little understood. We do know that the affective states produced through music are universal, similar across all cultures.

The term "music," of course, covers many activities—composing, performing, listening, singing, dancing. Some activities, such as performing, dancing, and singing, are clearly behavioral. Some, such as composing and listening, are clearly visceral and reflective. The musical experience can range from the one extreme where it is a deep, fully engrossing experience where the mind is fully immersed to the other extreme, where the music is played in the background and not consciously attended to. But even in the latter case, the automatic, visceral processing levels almost definitely register the melodic and rhythmic structure of the music, subtly, subconsciously, changing the affective state of the listener.

Music impacts all three levels of processing. The initial pleasure of the rhythm, tunes, and sounds is visceral, the enjoyment of playing and mastering the parts behavioral, and the pleasure of analyzing the intertwined, repeated, inverted, transformed melodic lines reflective. To the listener, the behavioral side is vicarious. The reflective appeal can come several ways. At one extreme, there is the deep appreciation of the structure of the piece, perhaps of the reference it makes to other pieces of music. This is the level of music appreciation exercised by the critic, the connoisseur, or the scholar. At the other extreme the musical structure and lyrics might be designed to delight, surprise, or shock.

Finally, music has an important behavioral component, either because the person is actively engaged in playing the music or equally actively singing or dancing. But someone who is just listening can also be behaviorally engaged by humming, tapping, or mentally following—and predicting—the piece. Some researchers believe that music is as much a motor activity as a perceptual one, even when simply listening. Moreover, the behavioral level could be involved vicariously, much as it is for the reader of a book or the viewer of a film (a topic I discuss later in this chapter).

Rhythm is built into human biology. There are numerous rhythmic patterns in the body, but the ones of particular interest are those that are relevant to the tempos of music: that is, from a few events per second to a few seconds per event. This is the range of body functions such as the beating of the heart and breathing. Perhaps more important, it is also the range of the natural frequencies of body movement, whether walking, throwing, or talking. It is easy to tap the limbs within this range of rates, hard to do it faster or slower. Much as the tempo of a clock is determined by the length of its pendulum, the body can adjust its natural tempo by tensing or relaxing muscles to adjust the effective length of the moving limbs, matching their natural rhythmic frequency to that of the music. It is therefore no accident that in playing music, the entire body keeps the rhythm.

All cultures have evolved musical scales, and although they differ, they all follow similar frameworks. The properties of octaves and of consonant and dissonant chords derive in part from physics, in part from the mechanical properties of the inner ear. Expectations play a central role in creating affective states, as a musical sequence satisfies or violates the expectations built up by its rhythm and tonal sequence. Minor keys have different emotional impact than major keys, universally signifying sadness or melancholy. The combination of key structure, choice of chords, rhythm, and tune, and the continual buildup of tension and instability create powerful affective influences upon us. Sometimes these influences are subconscious, as when music plays in the background during a film, but deliberately scored to invoke specific affective states. Sometimes these are conscious and deliberate, as when we devote our full conscious attention to the music, letting ourselves be carried vicariously by the impact, behaviorally by the rhythm, and reflectively as the mind builds upon the affective state to create true emotions.

We use music to fill the void when pursuing otherwise mindless activities, while stuck on a long, tiring trip, walking a long distance, exercising, or simply killing time. Once upon a time, music was not portable. Before the invention of the phonograph, music could be heard only when there were musicians. Today we carry our music players with us and we can listen twenty-four hours a day if we wish. Airlines realize music is so essential that they provide a choice of

FIGURE 4.7a and b

Music everywhere.

While drilling holes or recharging batteries, while taking photographs, on your cell phone. And of course, while driving your car, jogging, flying in an airplane, or just plain listening to music. Figure a shows the DEWALT battery charger for portable tools, with built-in radio; figure b shows an MP3 player built into a digital camera.

(Image a courtesy of DeWALT Industrial Tool Co. Image b *courtesy of Fujifilm USA. Note: This model is no longer available.)*

styles and hours of selections at every seat. Automobiles come equipped with radios and music players. And portable devices proliferate apparently endlessly, being either small and portable or combined with any other device the manufacturer thinks you might have with you: watches, jewelry, cell phones, cameras, and even work tools (figure 4.7a & b). Whenever I have had construction work done on a home, I noted that, first, the workers brought in their music players, which they set up in some central location with a super-loud output; then they would bring in their tools, equipment, and supplies. DEWALT, a manufacturer of cordless tools for construction workers, noticed the phenomenon and responded cleverly by building a radio into a battery charger, thus combining two essentials into one easy-to-carry box.

The proliferation of music speaks to the essential role it plays in our emotional lives. Rhyme, rhythm, and melody are fundamental to our emotions. Music also has its sensuous, sexual overtones, and for all these reasons, many political and religious groups have attempted to ban or regulate music and dance. Music acts as a subtle, subconscious enhancer of our emotional state throughout the day. This is why it is ever present, why it is so often played in the background in stores, offices, and homes. Each location gets a different style of music: Peppy, rousing beats would not be appropriate for most office work (or funeral homes). Sad, weepy music would not be conducive to efficient manufacturing.

The problem with music, however, is that it can also annoy—if it is too loud, if it intrudes, or if the mood it conveys conflicts with the listener's desires or mood. Background music is fine, as long as it stays in the background. Whenever it intrudes upon our thoughts, it ceases to be an enhancement and becomes an impediment, distracting, and irritating. Music must be used with delicacy. It can harm as much as help.

But if music can be annoying, what about the intrusive nature of today's beeping, buzzing, ringing electronic equipment? This is noise pollution gone rampant. If music is a source of positive affect, electronic sounds are a source of negative affect.

In the beginning was the beep. Engineers wanted to signal that some operation had been done, so, being engineers, they played a short tone. The result is that all of our equipment beeps at us. Annoying, universal beeps. Alas, all this beeping has given sound a bad name. Still sound, when used properly, is both emotionally satisfying and informationally rich.

Natural sounds are the best conveyers of meaning: a child laughing, an angry voice, the solid "clunk" when a well-made car door closes. The unsatisfying tinny sound when an ill-constructed door closes. The "kerplunk" when a stone falls into the water.

But so much of our electronic equipment now bleats forth unthinking, unmusical sounds that the result is a cacophony of irksome beeps or otherwise unsettling sounds, sometimes useful, but mostly emotionally upsetting, jarring, and annoying. When I work in my kitchen, the pleasurable activities of cutting and chopping, breading and sautéing, are continually disrupted by the dinging and beeping of timers, keypads, and other ill-conceived devices. If we are to have devices that signal their state, why not at least pay some attention to the aesthetics of the signal, making it melodic and warm rather than shrill and piercing?

It is possible to produce pleasant tones instead of irritating beeps. The kettle in figure 4.8 produces a graceful chord when the water boils. The designers of the Segway, a two-wheeled personal transporter, "were so obsessed with the details on the Segway HT that they designed the meshes in the gearbox to produce sounds exactly two musical octaves apart—when the Segway HT moves, it makes music, not noise."

Some products have managed to embed playfulness as well as information into their sounds. Thus, my Handspring Treo, a combined cellular telephone and personal digital assistant, has a pleasant three-note ascending melody when turned on, descending when turned off. This provides useful confirmation that the operation is being performed, but also a cheery little reminder that this pleasant device is obediently serving me.

Cell phone designers were perhaps the first to recognize that they could improve upon the grating artificial sounds of their devices. Some phones now produce rich, deep musical tones, allowing pleasant tunes to replace jarring rings. Moreover, the owner can select the

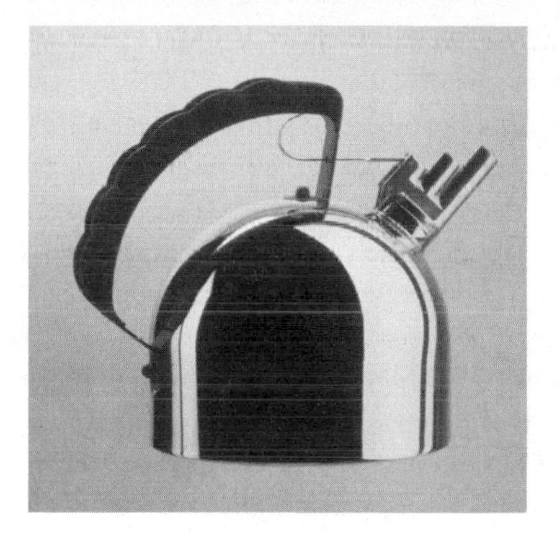

FIGURE 4.8

Richard Sapper's *kettle with singing whistle,* produced by Alessi. Considerable effort was given to the sound produced by the whistling spout: a chord of "e" and "b," or, as described by Alberto Alessi, "inspired by the sound of the steamers and barges that ply the Rhine."

(Alessi "9091," Design by Richard Sapper, 1983, Kettle with melodic whistle. Image courtesy of Alessi.)

sounds, allowing each individual caller to be associated with a unique sound. This is especially valuable with frequent callers and friends. "I always think of my friend when I hear this tune, so I made it play whenever he calls me," said one cell phone user to me, describing how he chose "ring tones" appropriate to the person who was calling: joyful pleasant tunes to joyful pleasant people; emotionally significant tunes for those who have shared experiences; sad or angry sounds to sad or angry people.

But even were we to replace the grating electronic tones with more pleasant musical sounds, the auditory dimension still has its drawbacks. On the one hand, there is no question that sound—both musical and otherwise—is a potent vehicle for expression, providing delight, emotional overtones, and even memory aids. On the other hand, sound propagates through space, reaching anyone within range equally, whether or not that person is interested in the activity: The musical ring that is so satisfying to a telephone's owner is a disturbing interruption to others within earshot. Eyelids allow us to shut out light; alas, we have no earlids.

When in public spaces—the streets of a city, in a public transit system, or even in the home—sounds intrude. The telephone is, of course, one of the worst offenders. As people speak loudly to make sure they are heard by their correspondent, they also cause themselves to be heard by everyone within range. Telephones, of course, are not the only intrusions. Radios and television sets, and the beeps and bongs of our equipment. More and more equipment comes equipped with noisy fans. Thus, the fans of heating and air-conditioning equipment can drown out conversation, and the fans of office equipment and home appliances add to the tensions of the day. When we are out of doors, we are bombarded by the sounds of passing aircraft, the horns and engine sounds of motor traffic, the warning back-up horns of trucks, the loud music players of others, emergency sirens, and the ever-present, shrill sounds of the cellular telephone ring, often mimicking a full musical performance. In public spaces, we are far too frequently interrupted by public announcements, starting with the completely unnecessary but annoying "Attention, Attention," followed by an announcement only of interest to a single person.

There is no excuse for this proliferation of sounds. Many cell phones have the option to set their rings to a private vibration, felt by the desired recipient but no others. Necessary sounds could be made melodious and pleasant, following the lead of the Sapper kettle in figure 4.8 or the Segway. Cooling and ventilation fans could be designed to be quiet as well as efficient by reducing their speed and increasing their blade size. The principles of noise reduction are well known, even if seldom followed. Whereas musical sounds at appropriate times and places are emotional enhancers, noise is a vast source of emotional stress. Unwanted, unpleasant sounds produce anxiety, elicit negative emotional states, and thereby reduce the effectiveness of all of us. Noise pollution is as negative to people's emotional lives as other forms of pollution are to the environment.

Sound can be playful, informative, fun, and emotionally inspiring. It can delight and inform. But it must be designed as carefully as any other aspect of design. Today, little thought is given to this side of design, and so the result is that the sounds of everyday things annoy many while pleasing few.

Seduction at the Movies

All the theatrical arts engage the viewer both cognitively and emotionally. As such, they are perfect vehicles to explore the dimensions of pleasure. In my research for this book, I discovered Jon Boorstin's analysis of films, a marvelous example of how the three levels of processing have their impact. His 1990 book, *The Hollywood Eye: What Makes Movies Work*, was such a wonderful fit to the analyses of my book that I just had to tell you about it.

Boorstin points out that movies appeal on three different emotional levels: *visceral*, *vicarious*, and *voyeur*, which bear perfect correspondence to my three levels of visceral, behavioral, and reflective. Let me start with the visceral side of movies. Boorstin's description of this component of a film is pretty much identical with my visceral level. Indeed, the match was so good, that I decided to use his term instead of "reactive," the term I use in my scientific publications. The phrase "reactive design" didn't quite capture the correct intention, but once I read Boorstin, it was obvious that the phrase "visceral design" was perfect, at least for this purpose. (But I still use "reactive" in my scientific publications.)

The passions aroused in film, says Boorstin, "are not lofty, they're the gut reactions of the lizard brain—the thrill of motion, the joy of destruction, lust, blood lust, terror, disgust. Sensations, you might say, rather than emotions. More complex feelings require the empathic response, but these simple, powerful urges reach up and grab us by the throat without an intermediary." He identifies the "slow-motion killing in *The Wild Bunch*, the monster in *The Fly*, or the bland titillation of soft-core porn" as examples of the visceral side of movies. Add the chase scene in *The French Connection* (or any classic spy or detective story), gun battles, fights, adventure stories, and, of course, horror and monster movies, and you have classic visceral level adventures.

Note the critical role played by music and lighting: dark, creepy scenes and dark, foreboding music. Minor keys for sad or unhappy, jubilant bouncy melodies for positive affect. Bright colors and bright lighting versus dark, gloomy colors and lights all exert their visceral influence. Camera angle, too, exerts its influence. Too far away, and the viewer is no longer experiencing but, instead, observing vicariously. Too close, and the image is too large for direct immediate impact. Film from above, and the people in the scene are diminished; film from below, and the actors are powerful, imposing. These operations all work on the subconscious level. We are usually unaware of the techniques used by directors and photographers to manipulate our emotions. The visceral level becomes completely absorbed in the sights and sounds. Any awareness of the technique would occur on the reflective level and would distract from the visceral experience. In fact, the only way to critique a film is by becoming detached, removed from visceral reaction and able to ponder the technique, the lights, the camera movements and angles. It is difficult to enjoy the film while analyzing it.

BOORSTIN'S "VICARIOUS" level corresponds to my "behavioral" level. The word "vicarious" is appropriate because the viewers are not directly engaging in the filmed activities but are, instead, watching and observing. If the film is well crafted, they are enjoying the activities vicariously, experiencing them as if they were participating. As Boorstin says, "The vicarious eye puts our heart in the actor's body: we feel what the actor feels, but we judge it for ourselves. Unlike relationships in life, here we can give ourselves up to other people in full confidence that we will always be in command."

If the visceral level grabs the viewer in the guts, driving automatic reactions, the vicarious level involves the viewer in the story and emotional line of the movie. Normally, the behavioral level of affect is invoked by a person's activities: it is the level of doing and acting. In the case of a film, the viewer is passive, sitting in a theater, experiencing the action vicariously. Nonetheless, the vicarious experience can play upon the same affective system.

Here is the power of storytelling, of the script, the actors, transporting viewers into the world of make-believe. This is "the willful suspension of disbelief" that the English poet Samuel Taylor Coleridge discussed as being essential for poetry. Here is where you get captured, caught up in the story, identifying with the situation and the characters. To be fully engrossed within a movie is to feel the world fade away, time seem to stop, and the body enter the transformed state that the social scientist Mihaly Csikszentmihalyi has labeled "flow."

Csikszentmihalyi's flow state is a special, detached state of consciousness, in which you are aware only of the moment, of the activity, and of the sheer enjoyment. It can occur in almost any activity: skilled tasks, sports, video games, board games, or any kind of mind-absorbing work. You can experience it in the theater, reading a book, or with intense problem solving.

The conditions required for flow to occur include lack of distractions and an activity paced precisely to match your skills, pushing you slightly above your capabilities. The level of difficulty has to be just at the edge of capability: too difficult and the task becomes frustrating; too easy and it becomes boring. The situation has to engage your entire conscious attention. This intense concentration causes outside distractions to fade away and the sense of time to disappear. It is intense, exhausting, productive, and exhilarating. It is no wonder that Csikszentmihalyi and his colleagues have spent considerable time exploring the phenomenon in its many manifestations.

The key to success of the vicarious level in film is the development and maintenance of the flow state. The pace has to be appropriate to avoid frustration or boredom. There can be no interruptions or distractions that might divert attention if one is to become truly captured by flow. Whenever we speak of films or other entertainment as "escapist," we are referring to the ability of the vicarious state and the behavioral level of affect to disengage people from the cares of life and transport them into some other world.

THE VOYEURISTIC level is that of the intellect, standing back to reflect and observe, to comment and think about an experience. Here is where the depth and complexity of characters, events, and the metaphors and analogies that a movie is meant to convey produce a deeper, richer meaning than is visible on the surface with the characters and story. "The voyeur's eye," says Boorstin, "is the mind's eye, not the heart's."

The word "voyeur" often is used to refer to observation of sensual or sexual subjects, which is not the meaning intended here. Boorstin explains that by the term "voyeur," he means "not the sexual kink but Webster's second definition of the word: the voyeur is the 'prying observer.' The voyeur's pleasure is the simple joy of seeing the new and the wonderful."

The voyeur's eye demands explanation—this is the level of cognition, of understanding and interpreting. As Boorstin points out, the vicarious experience can be profoundly moving, but the voyeur's eye, ever watching, ever thinking, is logical and reflective: "The voyeur in us is logical to a fault, impatient, picky, literal, but if properly respected it gives the special pleasures of the new and the clever, of a fresh place or crisply thought-out story." Of course, the voyeur can generate emotional suspense as well. It is the voyeur who knows that the wicked villain is hiding in wait for the hero, that the trap seems inescapable, and therefore that the hero is about to face death or, at the least, pain and torture. This level of excitement requires the thinking mind, and, of course, a clever director who plays upon those conjectures.

But, as Boorstin also points out, the voyeur can ruin a perfectly good movie by critiquing it:

> It can ruin the most dramatic moment with the most mundane concern: "Where are they?" "How did she get in the car?" "Where did the gun come from?" "Why don't they call the police?" "He's already used six shots—how come he's still firing?" "They'd never get there in time!" For a movie to work, the voyeur's eye must be pacified. For a movie to work brilliantly, the voyeur's eye must be entranced.

Voyeuristic movies are reflective movies, for example *2001: A Space Odyssey*, which, except for one lengthy visceral section, is mind-numbing in its intellectualism and almost exclusively a reflective experience. *Citizen Kane* is a fine example of both an entrancing story and a voyeur's delight.

JUST AS our experiences do not come neatly divided into unique categories of visceral, behavioral, or reflective, so films cannot be stuck neatly into one of three packages: visceral, or vicarious, or voyeuristic. Most experiences, and most films, cut across the boundaries.

The best products and the best films neatly balance all three forms of emotional impact. Although *The Magnificent Seven* is, as Boorstin puts it, "seven guys saving a town from bandits," if that were all there was to it, it would not have become such a classic. The film started life in 1954 in Japan as *The seven samurai (Shichinin no samurai)*, a film directed by Akira Kurosawa. In Japan, it was the story of seven samurai warriors hired to save a village from murderous thieves. It was redone as an American western in 1960 as *The Magnificent Seven* by John Sturges. Both films follow the same story line (both are excellent, although many movie buffs prefer the original). And both films successfully capture the viewer at all three modes, with engaging visceral spectacles, an engrossing story for the vicarious, and enough depth and hidden metaphorical allusions to content the reflective voyeur.

Sound, color, and lighting also play critical roles. In the best of cases, they heighten the experience without conscious awareness. Background music, on the face of it, is strange, for it is present even in so-called realistic movies, even though no music plays in our everyday world of real life. Purists scoff at the use of music, but omit it and a movie suffers. Music seems to modulate our affective system to enhance the experience at all levels of involvement: visceral, vicarious, and voyeuristic.

Lighting can intensify experience. Although most films today are shot in color, the director and photographer can dramatically impact the film by the style of lighting and color. Bright primary colors are the one extreme, with subdued pastels or dimly lit scenes another. The extreme case of color is the decision not to use it: to film in black and white. Although rarely used anymore, black and white can convey powerful dramatic impact, quite different from that possible with color. Here, the cinematographer can make skillful use of contrasts—light and dark, subtle grays—to convey an image's emotional tone.

The craft of filmmaking encompasses a wide variety of domains. All elements of a film make a difference: story line, pace and tempo, music, framing of the shots, editing, camera position and movement. All come together to form a cohesive, complex experience. A thorough analysis can, and has, filled many books.

All these effects work best, however, when they are unnoticed by the viewer. *The Man Who Wasn't There* (directed and written by the Coen brothers) was filmed in black and white. Roger Deakins, the cinematographer, stated that he wanted black and white rather than color so as not to distract from the story; unfortunately, he fell in love with the power of monochrome images. The film has wonderfully glorious shots, with high dark/light contrast, and in some places spectacular backlighting, all of which I noticed. This is a no-no in film: if you notice it, it is bad. Noticing takes place at the reflective (voyeur's) level, distracting you from that suspension of disbelief so essential to becoming fully captured by the flow at the behavioral (vicarious) level.

The story line and engrossing exposition of *The Man Who Wasn't There* enhanced the vicarious pleasure of the film, but noticing the photography caused the voyeuristic pleasure to interrupt with internal commentary ("How did he do that?" "Look at the magnificent lighting," and so on) and throw the vicarious pleasure off track. Yes, you should be able to go back afterward and marvel at how a film was done, but this should not intrude upon the experience itself.

Video Games

> Overslept, woke at 8:00. Only time for a quick coffee before the car pool arrives. The kitchen is disgusting, didn't clean up after last night's little party. Need a bath, but no time (the bathroom's flooded anyway from the broken sink I never got around to fixing). Got to work late and in horrible shape, was demoted as a result. Got home at 5:00, the repo man promptly showed up and repossessed my television because I forgot to pay my bills. My girlfriend won't speak to me because she saw me flirting with the neighbor last night.

Did you realize that this quotation is the description of a game? Not only does it feel like real life, but a bad life at that. Why would anyone think it was a game? Aren't games supposed to be fun? Well, not only is it a description of a game, it is a best-selling one called "The Sims." Will Wright, the designer and inventor of the Sims, explained that this was a typical day in the life of a game character, as developed by a beginning player.

The Sims is an interactive simulated-world game, otherwise known as a "God" game or sometimes "simulated life." The player acts like a god, creating characters, populating their world with houses, appliances, and activities. In this game, the player does not control what the game characters do. Instead, the player can only set up the environment and make high-level decisions. The characters control their own lives, although they have to live within the environment and high-level rules established by the player. The result is quite often not at all what their god intended them to do. The quotation is one example of a character unable to cope within the world its god created. But, says Wright, as the player's skill at creating worlds improves, the character might be able to spend the end of each day "sipping mint-juleps by the pool."

Wright explains the problem like this:

> The Sims is really just a game about life. Most people don't consciously realize how much strategic thinking goes into everyday, minute-to-minute living. We're so used to doing it that it submerges into our subconscious as a background task. But each decision you make (which door to go through? where to eat lunch? when to go to bed?) is calculated at some level to optimize something (time, happiness, comfort). This game takes that internal process and makes it external and visible. One of the first things players usually do in the game is to recreate their family, home, and friends. Then they're playing a game about themselves, sort of a strange, surreal mirror of their own lives.

Play is a common activity, engaged in by many animals and, of course, by us humans. Play serves many purposes. It probably is good practice for many of the skills required later in life. It helps children develop the mix of cooperation and competition required to live effectively in social groups. In animals, play helps establish their social dominance hierarchy. Games are more organized than play, usually with formal or, at least, agreed-upon rules, with some goal and usually some scoring mechanism. As a result, games tend to be competitive, with winners and losers.

Sports are even more formally organized than games, and at the professional level, are as much for the spectator as the player. As a result, an analysis of spectator sports is somewhat akin to that of movies, where the experience is vicarious and as a voyeur.

Of all the varieties of play, games, and sports, perhaps the most exciting new development is that of the video game. This is a new genre for entertainment: literature, film, game playing, sports, interactive novel, storytelling—all of these, but more besides.

Video games were once thought of as a mindless sport for teenage boys. No more. They are now played all around the world, including slightly more than half the population of the United States. They are played by everyone: from children to mature adults, with the average age of a player around thirty, and the gender difference evenly split between men and women. Video games come in many genres. In *The Medium of the Video Game*, Mark Wolf identifies forty-two different categories:

> Abstract, Adaptation, Adventure, Artificial Life, Board Games, Capturing, Card Games, Catching, Chase, Collecting, Combat, Demo, Diagnostic, Dodging, Driving, Educational, Escape, Fighting, Flying, Gambling, Interactive Movie, Management Simulation, Maze, Obstacle Course, Pencil-and-Paper Games, Pinball, Platform, Programming Games, Puzzle, Quiz, Racing, Role-Playing, Rhythm and Dance, Shoot 'Em Up, Simulation, Sports, Strategy, Table-Top Games, Target, Text Adventure, Training Simulation, and Utility.

Video games are a mixture of interactive fiction with entertainment. During the twenty-first century, they promise to evolve into radically different forms of entertainment, sport, training, and education. Many games are fairly elementary, simply putting a player in some role where fast reflexes—and sometimes great patience—are required to traverse a relatively fixed set of obstacles in order to move up the levels either to obtain a total game score or to accomplish some simple goal ("rescue the beleaguered princess and save her kingdom"). But wait. The story lines are getting ever-more complex and realistic, the demands upon the player more reflective and cognitive, less visceral and fast motor responses. The graphics and sound are getting so good that simulator games can be used for real training, whether flying an airplane, operating a railroad, or driving a race car or automobile. (The most elaborate video games are the full-motion airplane simulators used by the airlines that are so accurate that they enable pilots to be certified to fly passenger planes without ever flying the actual aircraft. But don't call these "games"; they are taken very seriously, and some of them can cost as much as the airplane itself.)

Today's sales of video games approach—and, in some cases, surpass—box-office receipts of movies. And we are still in the early days of video games. Imagine what they will be like in ten or twenty years. In an interactive game what happens in a story depends as much upon your actions as on the plot set up by the author (designer). Contrast this with a movie, where you have no control over the events. As a result, when experienced game players watch a movie, they miss this control, feeling as if they are "stuck watching a one-way plot." Moreover, the sense of involvement, the flow state, is much more intense in games than in most movies. In movies, you sit at a distance watching events unfold. In a video game, you are an active participant. You are part of the story, and it is happening to you, directly. As Verlyn Klinkenborg says, "what underlies it all is that visceral sense of having walked through a door into another universe."

The interactive, controlling part of video games is not necessarily superior to the more rigid, fixed format of books, theater, and film. Instead, we have different types of experiences, both of which are desirable. The fixed formats let master storytellers control the events, guiding you through the events in a carefully controlled sequence, very deliberately manipulating your thoughts and emotions until the climax and resolution. You surrender yourself quite voluntarily to this experience, both for the enjoyment and for the lessons that might be learned about life, society, and humanity. In a video game, you are an active participant, and as a result, the experience may vary from time to time—dull, boring, frustrating during one session; exciting, invigorating, rewarding during another. The lessons to be learned will vary depending upon the exact sequence of events that occurred and whether or not you were successful. Books and films clearly have a permanent role in society, as do games, video or otherwise.

Books, theater, movies, and games all occupy a fixed period of time: there is a beginning, then an end. This is not so of life. Sure, birth marks the beginning and death the end, but from your everyday perspective, life is ongoing. It continues even when you sleep or travel. Life cannot be escaped. When you go away, you return to find out what has transpired in your absence (during those moments when you were not in touch via messaging, email, or telephone). Video games are becoming like life.

Video games used to involve single individuals. This will always be a viable genre, but more and more, these games are involving groups, sometimes scattered across the world, communicating through computer networks. Some are on-line, real-time activities, such as sports, games, conversations, entertainment, music, and art. But some are environments, simulated worlds with people, families, households, and communities. In all of these, life goes on even when you, the player, are not present.

Some games have already tried to reach out toward their human players. If you, the player, create a family in a "god" game, nurturing your invented characters over an extended duration, perhaps months or even years, what happens when a family member needs help while you are asleep, or at work, school, or play? Well, if the crisis is severe enough, your game-family member will do just what a real family member would do: contact you by telephone, fax, email, or whatever form works. Someday one might even contact your friends, asking for help. So don't be surprised if a co-worker interrupts you in an important business meeting to say that your game character is in trouble: it is in urgent need of your assistance.

Yes, video games are an exciting new development in entertainment. But they may turn out to be far more than entertainment. The artificial may cease to be distinguishable from the real.

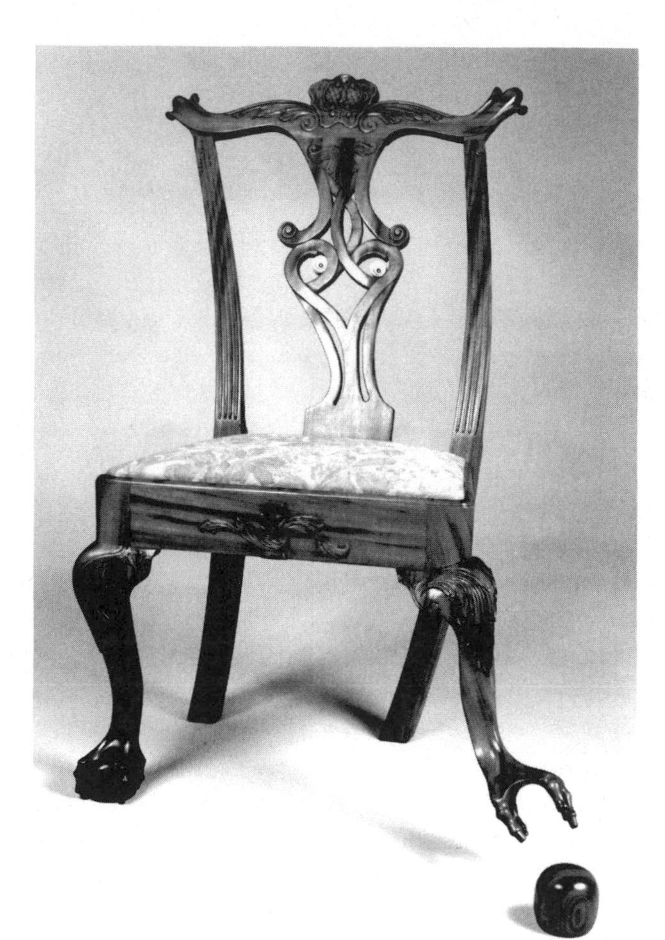

FIGURE 5.1

Oops! Uh oh, the poor chair.

It lost its ball, and it doesn't want anyone to know! Look how quietly it's sneaking out its foot, hoping to get it back before anyone notices.

(Renwick Gallery; image courtesy of Jake Cress, cabinetmaker.)

CHAPTER FIVE

People, Places, and Things

"UH OH, THE POOR CHAIR has lost its ball and doesn't want anyone to know." To me, the most interesting thing about the chair in figure 5.1 is that my reaction upon viewing "the poor chair" is perfectly sensible. Certainly I don't believe the chair is animate, that it has a brain, let alone feelings and beliefs. Yet there it is, clearly sneaking out its foot, hoping nobody will notice. What is going on?

This is an example of our tendency to read emotional responses into anything, animate or not. We are social creatures, biologically prepared to interact with others, and the nature of that interaction depends very much on our ability to understand another's mood. Facial expressions and body language are automatic, indirect results of our affective state, in part because affect is closely tied to behavior. Once the emotional system primes our muscles in preparation for action, other people can interpret our internal states by looking at how tense or relaxed we are, how our face changes, how our limbs move—

in short, our body language. Over millions of years, this ability to read others has become part of our biological heritage. As a result, we readily perceive emotional states in other people and, for that matter, anything that is at all vaguely lifelike. Hence our reaction to figure 5.1: the chair's posture is so compelling.

We have evolved to interpret even the most subtle of indicators. When we deal with people, this faculty is of huge value. It is even useful with animals. Thus, we can often interpret the affective state of animals—and they can interpret ours. This is possible because we share common origins for facial expression, gesture, and body posture. Similar interpretations of inanimate objects might seem bizarre, but the impulse comes from the same source—our automatic interpretive mechanisms. We interpret everything we experience, much of it in human terms. This is called anthropomorphism, the attribution of human motivations, beliefs, and feelings to animals and inanimate things. The more behavior something exhibits, the more we are apt to do this. We are anthropomorphic toward animals in general, especially our pets, and toward toys such as dolls, and anything we may interact with, such as computers, appliances, and automobiles. We treat tennis rackets, balls, and hand tools as animate beings, verbally praising them when they do a good job for us, blaming them when they refuse to perform as we had wished.

Byron Reeves and Clifford Nass have done numerous experiments that demonstrate—as the subtitle of their book puts it—"how people treat computers, television, and new media like real people and places." B. J. Fogg shows how people think of "computers as social actors," in his chapter of that title in his *Persuasive Technology*. Fogg proposes five primary social cues that people use to infer sociability with the person, or device, with whom, or which, they are interacting:

Physical: Face, eyes, body, movement

Psychological: Preferences, humor, personality, feelings, empathy, "I'm sorry"

Language: Interactive language use, spoken language, language recognition

Social Dynamics: Turn taking, cooperation, praise for good work, answering questions, reciprocity

Social Roles: Doctor, teammate, opponent, teacher, pet, guide

With the chair in figure 5.1, we succumb to the physical side. With computers, we often fall for the social dynamics (or, as is more often the case, the inept social dynamics). Basically, if something interacts with us, we interpret that interaction; the more responsive it is to us through its body actions, its language, its taking of turns, and its general responsiveness, the more we treat it like a social actor. This list applies to everything, human or animal, animate or non-animate.

Note that just as we infer the mental intentions of a chair without any real basis, we do the same for animals and other people. We don't have any more access to another person's mind than we do to the mind of an animal or chair. Our judgments of others are private interpretations based on observation and inference, not much different, really, than the evidence that makes us feel sorry for the poor chair. In fact, we don't have all that much information about the workings of our own minds. Only the reflective level is conscious: most of our motivations, beliefs, and feelings operate at the visceral and behavioral levels, below the level of awareness. The reflective level tries hard to make sense of the actions and behavior of the subconscious. But in fact, most of our behavior is subconscious and unknowable. Hence the need for others to aid us in times of trouble, for psychiatrists, psychologists, and analysts. Hence Sigmund Freud's historically impressive descriptions of the workings of id, ego, and superego.

So interpret we do, and over the many thousands or millions of years of evolution, we have coevolved muscle systems that display our emotions, and perceptual systems that interpret those of others. And with that interpretation also comes emotional judgment and empathy. We interpret, we emote. We can thereby believe that the object of our interpretations is sad or happy, angry or calm, sneaky or embarrassed. And, in turn, we ourselves can become emotional just by our interpretation of others. We cannot control those initial interpretations, for they come automatically, built in at the visceral level. We can control the final emotions through reflective analysis, but those initial impressions are subconscious and automatic. But, more important, it is this behavior that greases the wheels of social interaction, that makes it possible.

Designers take note. Humans are predisposed to anthropomorphize, to project human emotions and beliefs into anything. On the one hand, the anthropomorphic responses can bring great delight and pleasure to the user of a product. If everything works smoothly, fulfilling expectations, the affective system responds positively, bringing pleasure to the user. Similarly, if the design itself is elegant, beautiful, or perhaps playful and fun, once again the affective system reacts positively. In both cases, we attribute our pleasure to the product, so we praise it, and in extreme cases become emotionally attached to it. But when the behavior is frustrating, when the system appears to be recalcitrant, refusing to behave properly, the result is negative affect, anger, or worse, even rage. We blame the product. The principles for designing pleasurable, effective interaction between people and products are the very same ones that support pleasurable and effective interaction between individuals.

Blaming Inanimate Objects

> It starts out with slight annoyance, then the hairs on your neck start to prickle and your hands begin to sweat. Soon you are banging your computer or yelling at the screen, and you might well end up belting the person sitting next to you.
>
> —*Newspaper article on "Computer Rage"*

Many of us have experienced the computer rage described in the epigraph. Computers can indeed be infuriating. But why? And why do we get so angry at inanimate objects? The computer—or for that matter, any machine—doesn't intend to anger; machines have no intentions at all, at least not yet. We get angry because that's how our mind works. As far as we are concerned, we have done everything right, so the inappropriate behavior is therefore the fault of the computer. The "we" who faults the computer comes from the reflective level of our minds, the level that observes and passes judgment. Negative judgments lead to negative emotions, which can then inflame the judgments. The system for making judgments—cognition—is tightly coupled with the emotional system: each reinforces the other. The longer a problems lasts, the worse it becomes. Mild unhappiness is transformed into strong unhappiness. Unhappiness is transformed into anger, and anger into rage.

Note that when we get angry at our computer, we are assigning blame. Blame and its opposite, credit, are social judgments, assigning responsibility. This requires a more complex affective assessment than the dissatisfaction or pleasure one gets from a well- or ill-designed product. Blame or credit can come about only if we are treating the machine as if it were a causal agent, as if it made choices, in other words, as a human does.

How does this happen? Neither the visceral nor the behavioral level can determine causes. It is the role of reflection to understand, to interpret and find reasons, and to assign causes. Most of our rich, deepest emotions are ones where we have attributed a cause to an occurrence. These emotions originate from reflection. For example, two of the simpler emotions are hope and anxiety, hope resulting from expectation of a positive result, anxiety from expectation of something negative. If you are anxious, but the expected negative outcome doesn't happen, your emotion is one of relief. If you expect something positive, you are hopeful, and if it doesn't happen, then you feel disappointment.

So far, this is pretty simple, but suppose you—at your reflective level, to be more precise—decide that the result was someone's fault? Now we get into the complex emotions. Whose fault was it? When the result is negative and the blame put on yourself, you get remorse, self-anger, and shame. If you blame someone else, then you feel anger and reproach.

When the result is positive and the credit yours, you get pride and gratification. When the credit is someone else's, you get gratitude and admiration. Note how emotions reflect the interaction with others. Affect and emotion constitute a complex subject, involving all three levels, with the most complex emotions dependent upon just how the reflective level attributes causes. Reflection, therefore, is at the heart of the cognitive basis of emotions. The important point is that these emotions apply equally well to things as to people, and why not? Why distinguish between animate and inanimate things? You build up expectations of behavior based upon prior experience, and if the items with which you interact fail to live up to expectations, that is a violation of trust, for which you assign blame, which can soon lead to anger.

Cooperation relies on trust. For a team to work effectively each individual needs to be able to count on team members to behave as expected. Establishing trust is complex, but it involves, among other things, implicit and explicit promises, then clear attempts to deliver, and, moreover, evidence. When someone fails to deliver as expected, whether or not trust is violated depends upon the situation and upon where the blame falls.

Simple mechanical objects can be trusted, if only because their behavior is so simple that our expectations are apt to be accurate. Yes, a support or a knife blade may break unexpectedly, but that is about the largest possible transgression a simple object can do. Complex mechanical devices can go wrong in many more ways, and many a person has fallen in love—or become outraged—over the transgressions of automobiles, shop equipment, or other complex machinery.

When it comes to a lack of trust, the worst offenders of all are today's electronic devices, especially the computer (although the cell phone is rapidly gaining ground). The problem here is that you don't know what to expect. The manufacturers promise all sorts of wonderful results; but, in fact, the technology and its operations are invisible, mysteriously hidden from view, and often completely arbitrary, secretive, and sometimes even contradictory. Without any way of understanding how they operate or what actions they are doing, you can feel out of control and frequently disappointed. Trust eventually gives way to rage.

I believe that those of us who become angry with today's technology are justified. It may be an automatic result of our affective and emotional systems. It may not be rational, but so what? It is appropriate. Is it the computer's fault, or is it the software that runs within it? Is it really the software's fault, or is it the programmers who neglected to understand our real needs? As users of the technology, we don't care. All we care about is that our lives are made more frustrating. It is "their fault," "their" being everyone and everything involved in the computer's development. After all, these systems do not do a very good job of gathering trust. They lose files and they crash, oftentimes for no apparent reason. Moreover, they express no shame, no blame. They don't apologize or say they are sorry. Worse, they appear to blame us, the poor unwitting users. Who is "they"? Why does it matter? We are angered, and appropriately so.

Trust and Design

My 10-inch Wusthof chef knife. I could go on about the feel and aesthetic beauty, but upon further introspection I think my emotional attachment is substantially based on trust that comes from experience.

I know that my knife is up to whatever task I use it for. It is not going to slip out of my hand, the blade it not going to snap or break no matter how much pressure I apply; it is sharp enough to cut bones; it is not going to mutilate the meal I am about to serve to guests. I hate cooking in other people's kitchens and using their cutlery, even if it is good quality stuff.

This is a durable good, meaning I will only need to buy chef knives

> once or twice in a lifetime. I liked it OK when I purchased it, but my emotional attachment to it has developed over time through literally hundreds of consecutive positive experiences. This object is my friend.

The response above, one of many I received offering examples of products that people have learned to love or hate, vividly demonstrates the importance and power and properties of trust. Trust implies several qualities: reliance, confidence, and integrity. It means that one can count on a trusted system to perform precisely according to expectation. It implies integrity and, in a person, character. In artificial devices, trust means having it perform reliably, time after time after time. But there is more. In particular, we have high expectations of systems we trust: we expect them "to perform precisely according to expectation," which, of course, implies that we have built up particular expectations. These expectations come from multiple sources: the advertisements and recommendations that led us to buy the item in the first place; the reliability with which it has been performing since we got it; and, perhaps most important of all, the conceptual model we have of the item.

Your conceptual model of a product or service—and the feedback that you receive—is essential in establishing and maintaining trust. As I discussed in chapter 3, a conceptual model is your understanding of what an item is and how it works. If you have a good, accurate conceptual model, especially if the item keeps you informed about what it is doing—what stage in the operations it has reached, and whether things are proceeding smoothly or not—then you are not surprised by the result.

Consider what happens when your car runs out of gasoline. Whose fault is it? It depends. Most people's conceptual model of a car includes a fuel gauge that says what percentage of the tank is filled with gasoline. Many people also expect a warning such as a flashing light when the tank is close to empty. Some people even rely upon their assumption that the gas gauge is conservative, indicating that the tank is emptier than it really is, giving some leeway.

Suppose that the gas gauge has been reading close to empty, the warning light has flashed, but you procrastinate, not wanting to take the time to refill. If you run out of gasoline, you will blame yourself. Not only will you not be upset at the car, you might even now trust it more than ever. After all, it indicated you were going to run out of fuel, and you did. What if the warning light never came on? In that case you would blame the car. What if the gas gauge fluctuated up and down, continually varying? Then you wouldn't know how to interpret it: you wouldn't trust it.

Do you trust the gas gauge of your car? Most people are wary at first. When they drive in a new car, they have to do some tests to discover how much to trust the gas gauge. The typical way is to drive the car to lower and lower fuel estimates before refilling. The true test, of course, would be to run out of fuel deliberately in order to see how that corresponded to the meter reading, but most people don't need that much reassurance. Rather, they do enough driving to determine how much to trust the indicator, whether it be the meter reading or the low-fuel warning light in some cars, or for those with trip computers, the miles of driving the computer predicts can be done with the remaining fuel. With sufficient experience, people learn how to interpret the readings and, thus, how much to trust the gauge. Trust has to be earned.

Living in an Untrustworthy World

> It's human nature to trust our fellow man, especially when the request meets the test of being reasonable. Social engineers use this knowledge to exploit their victims and to achieve their goals.
>
> —*K. D. Mitnick and W. L. Simon,*
> The Art of Deception

Trust is an essential ingredient in cooperative, human interaction. Alas, this also makes it a vulnerability, ready for exploitation by what is called "social engineering," the crooks, thieves, and terrorists who exploit and manipulate our trust and good nature for their gain. As more and more of our everyday objects are manufactured with computer chips inside, with intelligence and flexibility, and with communication channels to the other devices in our environment and to the worldwide network of information and services, it is critical to worry about those who would do harm, whether by accident, for the sake of mischief, for fun, or with malicious intent to defraud or harm. Crooks, thieves, criminals, and terrorists are experts at exploiting the willingness of people to help one another, both to figure out how to use onerous technology and when someone appears to be in urgent need of assistance.

A common approach to improved safety and security is to tighten up on procedures and to require redundant checking. But as more people are involved in checking a task, safety can decrease. This is called "bystander apathy," a term that came from studies of the 1964 murder of Kitty Genovese on the streets of New York City. Although numerous people witnessed that incident, no one helped. At first the lack of response was simply blamed on the callousness of New York City residents, but social psychologists Bibb Latané and John Darley were able to repeat the bystander behavior, both in their laboratory and in field studies. They concluded that the more people watching, the less likely anyone would help. Why?

Think about your own reaction. If you were by yourself, walking along the streets of a large city and encountered what looked like a crime, you might be frightened and, therefore, reluctant to intervene. Still, you probably would try to call for help. But suppose a crowd of people were watching the incident? What would you do then? You probably would assume that you weren't witnessing anything serious, because if it were, people in the crowd would be doing something. The fact that nobody is doing anything must mean that nothing bad is happening. After all, in a large city, anything might happen: maybe it's actors making a movie.

Bystander apathy works in security as well. Suppose that you are working as a technician at a power plant. Among your jobs, you are supposed to check the meter readings with one of your colleagues, another technician at the plant, a person you know and trust. Moreover, when you have finished, your supervisor will also do a check. The result is that you don't have to exert extra care on the task. After all, how could a mistake get through with so many people? The problem is that everyone feels this way. As a result, the more people that check on something, the less carefully each person performs the task. As more people are responsible, security may diminish: trust gets in the way.

The commercial aviation community has done an excellent job of fighting this tendency with its program of "Crew Resource Management." All modern commercial aircraft have two pilots. One, the more senior, is the captain, who sits in the left-hand seat, while the other is the first-officer, who sits in the right-hand seat. Both are qualified pilots, however, and it is common for them to take turns piloting the aircraft. As a result, they are referred to by the terms "pilot flying" and "pilot not flying." A major component of crew resource management is that the pilot who is not flying be an active critic, continually checking and questioning the actions taken by the pilot who is flying. The pilot flying is supposed to thank the other for the questions, even when they are unnecessary, or even wrong. Obviously, getting this process in place was difficult, for it involved major changes in the culture, especially when one pilot was junior. After all, when one person questions another's behavior, it implies a lack of trust; and when two people are supposed to work together, especially when one is superior to the other, trust is essential. It took a while before the aviation community learned to take the questioning as a mark of respect, rather than a lack of trust, and for senior pilots to insist that junior ones question all of their actions. The result has been increased safety.

Criminals and terrorists take advantage of misplaced trust. One strategy to break into a well-guarded place is to trigger the alarms repeatedly over the course of a few days, and then hide so that the security personnel cannot find any cause for the trigger. Eventually, in frustration over the repeated false alarms, the security people will no longer trust them. It is then the criminals break in.

Not everyone is untrustworthy, just a few—but those few can be so severely disruptive that we have little choice but to relinquish trust and be suspicious of everyone, everything. There is a terrible tradeoff here: the very things that make security tighter are often those that make our lives more difficult or, in some cases, impossible. We need more realistic security that is cognizant of human behavior.

Security is more of a social or human problem than a technological one. Sure, put in all the technology you like. Those who wish to steal, corrupt, or disrupt will find a way to take advantage of human nature and bypass the security. Indeed, excessive technology gets in the way of security, because, by making the task of conscientious, everyday workers more difficult, it makes the job of bypassing the security measures even easier. When the security codes or procedures become too complex, people can't remember them, so they will write them down and post them on their computer terminals, under their keyboards or phones, or in their desk drawer (on top, though, where they are easy to get to).

As I was writing this book, I served on a committee of the United States National Research Council investigating information technology and counterterrorism. For my section of the report, I studied the social engineering practices used by terrorists, criminals, and other troublemakers. Actually, it's not difficult to find this information. The basic principles have been around for centuries and there are many books by ex-criminals, law-enforcement officers, and even guides to writing crime novels that provide relevant information. The internet makes the research easy.

Want to break into a secure facility? Walk up to the door carrying an armload of computers, parts, and dangling cords. Ask someone to hold open the door, and thank them. Carry the junk over to an empty cubicle, look for the password and login name, which will be posted somewhere, and log in (figure 5.2). If you can't log in, ask someone for help. Just ask. As one handbook that I found on the internet puts

FIGURE 5.2a and b

How not to safeguard a password.

Figure a shows a note posted on the side of the computer display; figure b is an enlargement of the note. This is the sort of behavior that social engineers count on. But it is bad password policies that make us have to resort to this. Even if the password wasn't attached to the computer, a good social engineer could have guessed it: this computer is at the corporate headquarters of a major manufacturer of office furniture. "Chair"? Who would ever guess? *(Photograph by author.)*

it: Just shout, "Does anyone remember the password for this terminal?" You would be surprised how many people will tell you.

In the end, security is a systems problem, where the human is the most important component. When security procedures get in the way of well-meaning, dedicated workers, they will find work-arounds to avoid disruption, thus defeating the whole point of the procedure. The very attributes that make us effective, cooperative, creative workers, able to adapt to the unexpected and to provide assistance to others, make us vulnerable to those who would take advantage of us.

• • •

Communications That Serve Emotion

Everywhere is nowhere. When a person spends all his time in foreign travel, he ends by having many acquaintances, but no friends.

—Lucius Annaeus Seneca (5 BC–AD 65)

In my consulting work, I am often called upon to predict the next "killer application," to discover the next product that will be so popular that everyone will have to own it. Unfortunately, if I have learned anything, it is that precise predictions of this sort are simply not possible. The field is littered with the bodies of those who have tried. Moreover, it is possible to be correct about a prediction, but very far off as to its time frame. I predict that automobiles will drive themselves. When? I have no idea: it might be twenty years, it might be one hundred. I predict that video telephones will become so popular that they will be everywhere, and we will simply take them for granted. In fact, people might complain if there weren't any video. When? Forecasters have been predicting widespread adoption of video phones "in just a few years" for the last fifty years. Even successful products can take decades before they catch on.

But even if exact prediction of successful products is not possible, we can be certain of one category that almost always guarantees success: social interaction. Throughout the last one hundred years, as technologies have changed, the importance of communication has remained high on the list of essentials. For individual communication, this has meant mail, the telephone, email, cell phones, and instant messaging and text messaging on computers and cell phones. For organizations, add the telegraph, the corporate memo and newsletter, the fax machine, and the intranet, that specialization of the internet for intracompany communication and interaction. And for societal groups, add the town crier, the daily newspaper, radio, and television.

Up to a few years ago, the increasing ease and lowered cost of travel had the unfortunate side effect of weakening the bonds that hold people together. Yes, through letters and telephone people could still be somewhat in touch, but this touch was limited. Two thousand years ago the Roman philosopher Seneca complained that travel led to many acquaintances but few friends, and up to recently this complaint still held true. Distance used to matter. Move away from family and friends, and the contact waned. Sure, one could use mail and telephone, but these were sparse communications amidst the busy activities of the day. People who separated physically would often separate socially and emotionally as well.

No more: today we can be in continual contact with friends and relatives no matter where we are, no matter the time of day. Today's technology makes it possible to stay in touch with friends and family on a continual basis. Email, instant messaging, text messages, and voice mail have no barriers in time or distance. Travel is relatively easy by auto, train, or airplane. The mail system reliably traverses the earth. The telephone is readily available and, with the cellular phone, always with us, always on. Email is ubiquitous. Billions of short messages are sent daily among the cell phones of the world. The isolation once imposed by distance and separation is no longer true. Today we can easily keep in touch with one another to an amount undreamed of earlier. Moreover, the communication revolution has barely begun: if it is so pervasive now, at the start of the twenty-first century, what will it be in one hundred years?

Most of the short text messages appear to be content-free. Among teenagers, they are apt to say: "What are you doing?"—or, in the highly abbreviated form they often take, "watrudoin"; "Where are you? (wru)"; "See you later (cul8r)." Among business people during the business day, they differ slightly: "Boring meeting"; "What are you doing?"; "Want a drink after work?" Occasionally, of course, they have real substance, as in business negotiations or in arranging meeting times or the details of a contract. But, on the whole, the point of the frequent messages is not information sharing; it is emotional connecting. They are ways of saying to one another, "I'm here," "you are there," "we still like each other." People need to communicate continually, for comfort, for reassurance.

The real advantage of text messaging is that it can be used while you are doing other things. As long as your hands are free and you can sneak an occasional look at the screen, you can send and receive messages: in class, at business meetings, or even while conversing with others. There seem to be no bounds. Stick the device in your shirt pocket. Then, when bored, or when that pleasant vibrating sensation on the chest signifies the arrival of a new message, take it out and peek. Read the latest words and surreptitiously type a reply, using two thumbs on the tiny keyboard. Surreptitiously, because this is probably taking place at a meeting, where you are supposed to be attending to the speaker.

The ability to use short text messages so effortlessly has become a strong, emotional component of many people's lives. Numerous people who responded to my Internet request for experiences of bonding used the opportunity to tell me about their attachment to Instant Messaging (IM). Here are two responses:

> Instant messenger (IM) is an integrated part of my life. With it I have a sense of connection to many of my friends and colleagues around the world. Without it, I feel as though a window to part of my world is bolted shut.
>
> Another example is IM. I am so attached to my IM at work. I can't imagine my life without it. The real power of IM isn't the message (though that is a key attribute), but it's the presence detection. Knowing that someone "is there." Imagine knowing that every time you pick up the phone to dial someone there is going to be a real person to answer, and the person you want. That is the power of instant messaging.

The cell phone shares much of the emotional power of text messaging. It is much more than a simple communication device. Oh, sure, business thinks of it as a way of keeping in touch, of getting critical information to people when they need it, but that misses the whole point of these devices. It is fundamentally an emotional tool and a social facilitator. It keeps people in touch with one another. It lets friends chat: even if the formal, reflective content is vague, the emotional content is high. But although it lets us all share thoughts and ideas, music, and pictures, what it really lets us share is emotion. The ability to keep in touch throughout the day maintains a relationship, whether it be business or social.

Speech is a powerful social and emotional vehicle because it enables communication of emotional state through its natural prosody—pauses, rhythm, pitch inflections, hesitations, and repeats. Although text messaging is not as effective as speech at communicating emotion, it is superior as a tool for communication because it is unobtrusive. It can be kept private and it can be done secretly. I am always amused at business meetings by the sneaky, but skillful, use of text messaging. I watch otherwise serious, staid executives glance furtively down at their laps so as to read screens and type responses, all the while pretending to be listening to the meeting. Text messaging lets friends keep in touch, even when they should be attending to something else.

Isn't it strange that, although telephone service is an emotional tool, the appliance itself is not? People love the power of cell phone interaction, but do not seem to love any of the devices that make it possible. As a result, the turnover of devices is high. There is no product loyalty, no commitment to company or service provider. The cell phone, one of the most fundamentally emotional services, garners little attachment to its products.

Vernor Vinge, one of my favorite science fiction writers, wrote *A Fire Upon the Deep,* in which the planet Tines is populated by animals with a collective intelligence. These doglike creatures travel in packs, whose members are in continuous acoustical communication with one another, giving rise to a powerful, distributed consciousness. Individuals leave the pack because of death, illness, or accident, and new, young members are recruited to replace them, so that the pack maintains its identity far beyond that of any single individual. Each individual member of a pack lacks intelligence when all alone: the pack gains its intelligence through the collaboration of the many individuals. As a result, if an individual strays too far from the pack, the communication path is lost—for sound has limited range—and the resulting "singleton" is devoid of intelligence. Singletons rarely survive, and those that do are doomed to a mindless existence—literally mindless.

Walk down the street of any large city in any country of the world and watch the people who are talking on their cell phones: they are in their own space, physically adjacent to one location and one set of people but emotionally somewhere else. It is as if they fear being singletons in the crowd of strangers and opt instead to maintain connection with their pack, even if the pack is elsewhere. The cell phone establishes its own private space, removed from the street. Were the two people together, walking down the street, they would not be so isolated, for they would both be aware of one another, of the conversation and of the street. But with the cell phone, you enter into a private place that is virtual, not real, one removed from the surrounds, the better to bond with the other person and the conversation. And so you are lost to the street even while walking along it. Truly a private space in a public place.

Always Connected, Always Distracted

I HAVE watched phones ring and be answered in the most amazing places. In the movie theater, in the middle of board meetings. I once attended a meeting at the Vatican where I was part of a scientific delegation presenting our findings to the Pope. Cell phones were everywhere: each cardinal wore a gold chain upon which was hung a gold cross, each bishop had a gold chain upon which was hung a silver cross, but the head usher, who seemed to be the real person in charge, wore a gold chain upon which was hung a cell phone. The Pope may have been the center of attention, but I heard cell phones ringing continually throughout the ceremony. "Scusi," they would whisper into their phones, "I can't talk right now, I'm listening to the Pope."

On another occasion I was a member of a discussion panel in front of a large audience, when the moderator's cell phone rang just as he was in the middle of asking a question of one of the panelists. Yes, he answered it, disconcerting the panel members, but amusing the audience.

HURRAH FOR the communication technologies that allow us continuous contact with our colleagues, friends, and families, no matter where we are, no matter what we are doing! But, however powerful text and voice messages, phone calls, and emails are as tools for maintaining relationships or supervising work, note that one person's "keeping in touch" is another person's interruption. The emotional impact reflects this discrepancy: positive to the person keeping in touch, but negative and disturbing to the person being subjected to interruption.

There is a lack of symmetry in the perceived impact of an interruption. When I have lunch with friends who spend a considerable fraction of our time responding to calls on their cell phones, I consider this a distraction and an interruption. From their point of view, they are still with me, but the calls are essential to their lives and emotions and not at all an interruption. To the person taking the call, the time is filled, with information being conveyed. To me, it is empty unfilled time. The lunchtime conversation is now on hold. I have to wait for the interruption to end.

How much time does the interruption seem to take? To the person being interrupted, forever. To the person taking the call, just a few seconds. Perception is everything. When one is busy, times flies quickly. When there is nothing to do, it seems to drag. As a result, the person engaged in the cell phone conversation feels emotionally satisfied, while the other feels ignored and distanced, emotionally upset.

Human conscious attention is a component of the reflective level of the mind. It has limited capability. On the one hand, it limits the focus of consciousness, primarily to a single task. On the other hand, attention is readily distracted by changes in the environment. The result of this natural distractibility is a short attention span: new events continually engage attention. Today it is customary to argue that short attention spans are caused by advertisements, video games, music videos, and so on. But, in fact, the ready distractibility of attention is a biological necessity, developed through millions of years of evolution as a protective mechanism against unexpected danger: this is the primary function of the visceral level. This is probably why one byproduct of the negative affect and anxiety that results from perceived danger is a narrowing and focusing of attention. In danger, attention must not become distracted. But in the absence of anxiety, people are easily distracted, continually shifting attention. William James, the famous philosopher/psychologist, once said that his attention span was approximately ten seconds, and this in the late 1800s, far before the advent of modern distractions.

We carve out our own private spaces where needed. At home in our private study or bedroom, door locked if need be. At the office, in a private room or, struggling to accomplish privacy, in cubicles or shared space. In the library, helped by the no-talking rules and convention, or by private carrels for the privileged few. In streets, where people will gather to form clusters of conversations, seemingly oblivious to those around them, if only temporarily.

But the real problems of modern communication come from the limitations of human attention.

The limits on conscious attention are severe. When you are on a telephone call, you are doing a very special sort of activity, for you are a part of two different spaces, one where you are located physically, the other a mental space, the private location within your mind where you interact with the person on the other end of the conversation. This mental partitioning of space is a very special facility and it makes the telephone conversation, unlike other joint activities, demand a special kind of mental concentration. The result is that you are partially away from the real, physical space, even as you inhabit it. This division into multiple spaces has important consequences for the human ability to function.

Do you talk on a cell phone while driving an automobile? If so, you are dividing your conscious attention in a dangerous fashion, reducing your capacity to plan and anticipate. Yes, your visceral and behavioral levels of processing still function well, but not the reflective, the home of planning and anticipating. So driving is still possible, but primarily through the automatic, subconscious visceral and behavioral mechanisms. The part of driving that suffers is the reflective oversight, the planning, the ability to anticipate the actions of other drivers and any special conditions of the environment. That you can still appear to drive normally blinds you to the fact that the driving is less skillful, less able to cope with unexpected situations. Thus, the driving becomes dangerous, the cause being that distracting mental space. The danger does not stem from any requirement to hold the cell phone to your ear with one hand while steering with the other: a hands-free cell phone, where speaker and microphone are fixed to the car so that no hands are required, does not eliminate the distracting mental space. This is a new area of research, but the early studies seem to show that hands-free phones are just as dangerous as hand-held ones. The decrease in driver performance results from the conversation, not the telephone instrument.

Drive a car and hold a conversation with the passengers and, yes, some of the same distraction is present, especially because our social nature means that we tend to look at the person with whom we are conversing. Once again, the safety research is still at an early stage, but I predict that a conversation with real passengers will prove to be not as dangerous as one with far-away people, for the mental space we create for real passengers includes the auto and its surround, whereas the other one distances us from the auto. After all, we evolved to interact with others in the midst of other activities, but the evolutionary process could not anticipate communication at a distance.

We cannot take part in two intense conversations at the same time, at least not without degrading the quality and speed of each. Of course, we can and do take part in multiple instant- and text-messaging conversations "simultaneously," but the quote marks around "simultaneously" reflect that we don't really do the operations at the same instant, but rather interweave the two. Conscious, reflective attention is only necessary for reading and for formulation of new messages, but once formulated, the automatic behavioral level mechanisms can guide the actual keystrokes while reflection switches over to the other conversation.

Because most activities do not require continual, full-time conscious attention, we are able to go about our daily activities continually dividing attention among multiple diversions. The virtue of this division of attention is that we keep in touch with the environment: we are continually aware of the things around us. Walking down the street chatting with a friend, we still have considerable resources left for a multitude of activities: to notice the new stores that have opened on the block; to glance at the newspaper headlines; even to eavesdrop upon neighboring conversations. The difficulties arise only when we are forced to engage in mechanical activities, such as driving an automobile, where the technological demands can require immediate response. Here is where the apparent ease with which we can often do these tasks misleads us into thinking that full attention is never required. Our ability to handle distractions and to divide attention is essential to social interaction. Our ability to time-share, to do multiple activities, enhances these interactions. We are aware of others around us. We keep in touch with a large number of people. The continual switching of attention is normally a virtue, especially in the world of social interaction. In the mechanical world, it can be a peril.

By continually being in communication with friends across a lifetime, across the world, we risk the paradox of enhancing shallow interactions at the expense of deep ones. Yes, we can hold continual, short interactions with numerous people, thus keeping friendships alive. But the more we hold short, brief, fleeting interactions and allow ourselves to interrupt ongoing conversations and interactions, the less we allow any depth of interaction, any depth to a relationship.

"Continuously divided attention" is the way Linda Stone has described this phenomenon, but no matter how we may deplore it, it has become a commonplace aspect of everyday life.

The Role of Design

Technology often forces us into situations where we can't live without the technology even though we may actively dislike its impact. Or we may love what the technology provides us while hating the frustrations encountered while trying to use it. Love and hate: two conflicting emotions, but commonly combined to form an enduring, if uncomfortable, relationship. These love-hate relationships can be amazingly stable.

Love-hate relationships offer promise, if only the hate can be dissipated, retaining the love. The designer has some power here, but only to a limited extent, for although some of the irritation and dislike is a result of inappropriate or impoverished design, much is a result of societal norms and standards, and these can only be changed by society itself.

Much of modern technology is really the technology of social interaction: it is the technology of trust and emotional bonds. But neither social interaction nor trust were designed into the technology or even thought through: they came about through happenstance, through the accidental byproducts of deployment. To the technologist, the technology provides a means of communication; for us, however, it provides a means for social interaction.

There is much that can be done to enhance these technologies. We have already seen that lack of trust comes about from lack of understanding, from situations where we feel out of control, unaware of what has happened, or why, or what we should do the next time. And we have seen how the unscrupulous, the thief and the terrorist can take advantage of the normal trust people have for one another, a trust that is essential if normal civilization is to exist.

In the case of the personal computer, the frustrations and irritations that lead to "computer rage" are indeed the domain of design. These are caused by design flaws that exacerbate the problems. Some have to do with the lack of reliability and bad programming, some with the lack of understanding of human needs, and some the lack of fit between the operation of the computer and the tasks that people wish to do. All these can be solved. Today, communication seems always to be with us, whether we wish it to or not. Whether at work or play, school or home, we can make contact with others. Moreover, the distinctions among the various media are disappearing, as we send voice and text, words and images, music and video back and forth with increasing ease and frequency. When my friend in Japan uses his cell phone to take a photograph of his new grandchild and sends it to me in the United States, is this email, photography, or telephony?

The good news is that the new technologies enable us always to feel connected, to be able to share our thoughts and feelings no matter where we are, no matter what we are doing, independent of the time or time zone. The bad news is, of course, those very same things. If all my friends were always to keep in touch, there would be no time for anything else. Life would be filled with interruptions, twenty-four hours a day. Each interaction alone would be pleasant and rewarding, but the total impact would be overwhelming.

The problem, however, is that the ease of short, brief communication with friends around the world disrupts the normal, everyday social interaction. Here, the only hope is for a change in social acceptance. This can go in two directions. We could all come to accept the interruptions as a part of life, thinking nothing of it when the several members of a group continually enter their own private space to interact with others—friends, bosses, coworkers, family, or perhaps their video game, where their characters are in desperate need of help. The other direction is for people to learn to limit their interactions, to let the telephone take messages by text, video or voice, so that the calls can be returned at a convenient time. I can imagine solutions designed to help facilitate this, so that the technology within a telephone might negotiate with a caller, checking the calendars of each party and setting up a time to converse, all without bothering any of the individuals.

We need technologies that provide the rich power of interaction without the disruption: we need to regain control over our lives. Control, in fact, seems to be the common theme, whether it be to avoid the frustration, alienation, and anger we feel toward today's technologies, or to permit us to interact with others reliably, or to keep tight the bonds between us and our family, friends, and colleagues.

Not every interaction has to be done in real time, with participants interrupting one another, always available, always responding. The store-and-forward technologies—for example email and voice mail—allow messages to be sent at the sender's convenience, but then listened to or viewed at the receiver's convenience. We need ways of intermixing the separate communication methods, so that we could choose mail, email, telephone, voice, or text as the occasion demands. People need also to set aside time when they can concentrate without interruption, so that they can stay focused.

Most of us already do this. We turn off our cell phones and deliberately do not carry them at times. We screen our telephone calls, not answering unless we see—or hear—that it is from someone we really wish to speak to. We go away to private locations, the better to write, think, or simply relax.

Today, the technologies are struggling to ensure their ubiquitous presence, so that no matter where we are, no matter what we are doing, they are available. That is fine, as long as the choice of whether to use them remains with the individual at the receiving end. I have great faith in society. I believe we will come to a sensible accommodation with these technologies. In the early years of any technology, the potential applications are matched by the all-too-apparent drawbacks, yielding the love-hate relationship so common with new technologies. Love for the potential, hate for the actuality. But with time, with improved design of both the technology and the manner in which it is used, it is possible to minimize the hate and transform the relationship to one of love.

CHAPTER SIX

Emotional Machines

> Dave, stop . . . Stop, will you . . . Stop, Dave . . . Will you stop, Dave . . . Stop, Dave . . . I'm afraid. I'm afraid . . . I'm afraid, Dave . . . Dave . . . My mind is going . . . I can feel it . . . I can feel it . . . My mind is going . . . There is no question about it . . . I can feel it . . . I can feel it . . . I'm a. . . fraid.
>
> —*HAL, the all-powerful computer, in the movie* 2001

HAL IS CORRECT TO BE AFRAID: Dave is about to shut him down by dismantling his parts. Of course, Dave is afraid, too: HAL has killed all the other crew of the spacecraft and made an unsuccessful attempt on Dave's life.

But why and how does HAL have fear? Is it real fear? I suspect not. HAL correctly diagnoses Dave's intent: Dave wants to kill him. So fear—being afraid—is a logical response to the situation. But human emotions have more than a logical, rational component; they are tightly coupled to behavior and feelings. Were HAL a human, he would fight hard to prevent his death, slam some doors, do something to survive. He could threaten, "Kill me and you will die, too, as soon as the air in your backpack runs out." But HAL doesn't do any of this; he simply states, as a fact, "I'm afraid." HAL has an intellectual knowledge of what it means to be afraid, but it isn't coupled to feelings or to action: it isn't real emotion.

But why would HAL need real emotions to function? Our machines today don't need emotions. Yes, they have a reasonable amount of intelligence. But emotions? Nope. But future machines will need emotions for the same reasons people do: The human emotional system plays an essential role in survival, social interaction and cooperation, and learning. Machines will need a form of emotion—machine emotion—when they face the same conditions, when they must operate continuously without any assistance from people in the complex, ever-changing world where new situations continually arise. As machines become more and more capable, taking on many of our roles, designers face the complex task of deciding just how they shall be constructed, just how they will interact with one another and with people. Thus, for the same reason that animals and people have emotions, I believe that machines will also need them. They won't be human emotions, mind you, but rather emotions that fit the needs of the machines themselves.

Robots already exist. Most are fairly simple automated arms and tools in factories, but they are increasing in power and capabilities, branching out to a much wider array of activities and places. Some do useful jobs, as do the lawn-mowing and vacuum-cleaning robots that already exist. Some, such as the surrogate pets, are playful. Some simple robots are being used for dangerous jobs, such as fire fighting, search-and-rescue missions, or for military purposes. Some robots even deliver mail, dispense medicine, and take on other relatively simple tasks. As robots become more advanced, they will need only the simplest of emotions, starting with such practical ones as visceral-like fear of heights or concern about bumping into things. Robot pets will have playful, engaging personalities. With time, as these robots gain in capability, they will come to possess full-fledged emotions: fear and

FIGURE 6.1

C3PO (left) and R2D2 (right) of *Star Wars* fame.

Both are remarkably expressive despite R2D2's lack of body and facial structure.

(Courtesy of Lucasfilm Ltd.)

anxiety when in dangerous situations, pleasure when accomplishing a desired goal, pride in the quality of their work, and subservience and obedience to their owners. Because many of these robots will work in the home environment, interacting with people and other household robots, they will need to display their emotions, to have something analogous to facial expressions and body language.

Facial expressions and body language are part of the "system image" of a robot, allowing the people with whom it interacts to have a better conceptual model of its operation. When we interact with other people, their facial expressions and body language let us know if they understand us, if they are puzzled, and if they are in agreement. We can tell when people are having difficulty by their expressions. The same sort of nonverbal feedback will be invaluable when we interact with robots: Do the robots understand their instructions?

When are they working hard at the task? When are they being successful? When are they having difficulties? Emotional expressions will let us know their motivations and desires, their accomplishments and frustrations, and thus will increase our satisfaction and understanding of the robots: we will be able to tell what they are capable of and what they aren't.

Finding the right mix of emotions and intelligence is not easy. The two robots from the *Star Wars* films, R2D2 and C3PO, act like machines we might enjoy having around the house. I suspect that part of their charm is in the way they display their limitations. C3PO is a clumsy, well-meaning oaf, pretty incompetent at all tasks except the one for which he is a specialist: translating languages and machine communication. R2D2 is designed for interacting with other machines and has limited physical capabilities. It has to rely upon C3PO to talk with people.

R2D2 and C3PO show their emotions well, letting the screen characters—and the movie audience—understand, empathize with, and, at times, get annoyed with them. C3PO has a humanlike form, so he can show facial expressions and body motions: he does a lot of hand wringing and body swaying. R2D2 is more limited, but nonetheless very expressive, showing how able we are to impute emotions when all we can see is a head shaking, the body moving back and forth, and some cute but unintelligible sounds. Through the skills of the moviemakers, the conceptual models underlying R2D2 and C3PO are quite visible. Thus, people always have pretty accurate understanding of their strengths and weaknesses, which make them enjoyable and effective.

Movie robots haven't always fared well. Notice what happened to two movie robots: HAL, of the movie *2001* and David, of the movie *AI* (*Artificial Intelligence*). HAL is afraid, as the opening quotation of this chapter illustrates, and properly so. He is being taken apart—basically, being murdered.

David is a robot built to be a surrogate child, taking the place of a real child in a household. David is sophisticated, but a little too perfect. According to the story, David is the very first robot to have "unconditional love." But this is not a true love. Perhaps because it is "unconditional," it seems artificial, overly strong, and unaccompanied by the normal human array of emotional states. Normal children may love their parents, but they also go through stages of dislike, anger, envy, disgust, and just plain indifference toward them. David does not exhibit any of these feelings. David's pure love means a happy devoted child, following his mother's footsteps, quite literally, every second of the day. This behavior is so irritating that he is finally abandoned by his foster mother, left in the wilderness, and told not to come back.

The role of emotion in advanced intelligence is a standard theme of science fiction. Thus, two of the characters from the *Star Trek* television shows and films wrestle with the role of emotion and intelligence. The first, Spock, whose mother is human but whose father is Vulcan, has essentially no emotions, giving the story writers wonderful opportunities to pit Spock's pure reason against Captain Kirk's human emotions. Similarly, in the later series, Lieutenant Commander Data is pure android, completely artificial, and his lack of emotion provides similar fodder for the writers, although several episodes tinker with the possibility of adding an "emotion chip" into Data, as if emotion were a separate section of the brain that could be added or subtracted at will. But although the series is fiction, the writers did their homework well: their portrayal of the role of emotion in decision making and social interaction is reasonable enough that the psychologists Robert Sekuler and Randolph Blake found them excellent examples of the phenomena, appropriate for teaching introductory psychology. In their book, *Star Trek on the Brain*, they used numerous examples from the *Star Trek* series to illustrate the role of emotion in behavior (among other topics).

Emotional Things

How will my toaster ever get better, making toast the way I prefer, unless it has some pride? Machines will not be smart and sensible until they have both intelligence and emotions. Emotions enable us to translate intelligence into action.

Without pride in the quality of our actions, why would we endeavor to do better? The positive emotions are of critical importance to learning, to maintaining our curiosity about the world. Negative emotions may keep us from danger, but it is positive emotions that make living worthwhile, that guide us to the good things in life, that reward our successes, and that make us strive to be better.

Pure reason doesn't always suffice. What happens when there isn't enough information? How do we decide which course of action to take when there is risk, so that the possibility of harm has to be balanced against the emotional gain of success? This is where emotions play a critical role and where humans who have had neurological damage to their emotional systems falter. In the movie *2001*, the astronaut Dave risks his life in order to recover the dead body of his fellow astronaut. Logically this doesn't make much sense, but in terms of a long history of human society, it is of great importance. Indeed, this tendency of people to risk many lives in the effort to rescue a few—or even to retrieve the dead—is a constant theme in both our real lives and our fictional ones, in literature, theater, and film.

Robots will need something akin to emotion in order to make these complex decisions. Will that walkway hold the robot's weight? Is there some danger lurking behind the post? These decisions require going beyond perceptual information to use experience and general knowledge to make inferences about the world and then to use the emotional system to help assess the situation and move toward action. With just pure logic, we could spend all day frozen in place, unable to move as we think through all the possible things that could go wrong—as happens to some emotionally impaired people. To make these decisions we need emotions: robots will, too.

Rich, layered systems of affect akin to that of people are not yet a part of our machines, but some day they will be. Mind you, the affect required is not necessarily a copy of humans'. Instead, what is needed is an affective system tuned to the needs of the system. Robots should be concerned about dangers that might befall them, many of which are common to people and animals, and some of which are unique to robots. They need to avoid falling down stairs or off of other edges, so they should have fear of heights. They should get fatigued, so they do not wear themselves out and become low on energy (hungry?) before recharging their battery. They don't have to eat or use a toilet, but they do need to be serviced periodically: oil their joints, replace worn parts, and so on. They don't have to worry about cleanliness and sanitation, but they do need to pay attention to dirt that might get into their moving parts, dust and dirt on their television lenses, and computer viruses that might interfere with their functioning. The affect that robots require will be both similar to and very different from that of people.

Even though machine designers may have never considered that they were building affect or emotion into their machines, they have built in safety and survival systems. Some of these are like the visceral level of people: simple, fast-acting circuits that detect possible danger and react accordingly. In other words, survival has already been a part of most machine design. Many devices have fuses, so that if they suddenly draw more electric current than normal, the fuse or circuit breaker opens the electrical circuit, preventing the machine from damaging itself (and, along the way, preventing it from damaging us or the environment). Similarly, some computers have non-interruptible power supplies, so that if the electric power fails, they quickly and immediately switch to battery power. The battery gives them time to shut down in an orderly and graceful fashion, saving all their data and sending notices to human operators. Some equipment has temperature or water-level sensors. Some detect the presence of people and refuse to operate whenever someone is in a proscribed zone. Existing robots and other mobile systems already have sensors and visual systems that prevent them from hitting people and other objects or falling down stairs. So simple safety and survival is already a part of many designs.

In people and animals, the impact of the visceral system doesn't cease with an initial response. The visceral level signals higher levels of processing to try to determine the causes of the problem and to determine an effective response. Machines should do the same.

Any system that is autonomous—that is intended to exist by itself, without a caretaker always guiding it—continually has to decide which of many possible activities to do. In technical terms, it needs a scheduling system. Even people have difficulty with this task. If we are working hard to finish an important task, when should we take a break to eat, sleep, or to do some other activity that is perhaps required of us but not nearly so urgent? How do we fit the many activities that have to be done into the limited hours of the day, knowing when to put one aside, when not to? Which is more important: The critical proposal due tomorrow morning or planning a family birthday celebration? These are difficult problems that no machine today can even contemplate but that people face every day. These are precisely the sorts of decision-making and control problems for which the emotional system is so helpful.

Many machines are designed to work even though individual components may fail. This behavior is critical in safety-related systems, such as airplanes and nuclear power reactors, and very valuable in systems that are performing critical operations, such as some computer systems, hospitals, and anything dealing with the vital infrastructure of society. But what happens when a component fails and the automatic backups take over? Here is where the affective system would be useful.

The component failure should be detected at the visceral level and used to trigger an alert: in essence, the system would become "anxious." The result of this increased anxiety should be to cause the machine to act more conservatively, perhaps slowing down or postponing non-critical jobs. In other words, why shouldn't machines behave like people who have become anxious? They would be cautious even while attempting to remove the cause of anxiety. With people, behavior becomes more focused until the cause and an appropriate response are determined. Whatever the response for machine systems, some change in normal behavior is required.

Animals and humans have developed sophisticated mechanisms for surviving in an unpredictable, dynamic world, coupling the appraisals and evaluations of affect to methods for modulating the overall system. The result is increased robustness and error tolerance. Our artificial systems would do well to learn from their example.

Emotional Robots

> The 1980s was the decade of the PC, the 90s of the Internet, but I believe the decade just starting will be the decade of the robot.
>
> —*Sony Corporation Executive*

Suppose we wish to build a robot capable of living in the home, wandering about, fitting comfortably into the family—what would it do? When asked this question, most people first think of handing over their daily chores. The robot should be a servant, cleaning the house, taking care of the chores. Everyone seems to want a robot that will do the dishes or the laundry. Actually, today's dishwashers and clothes washers and dryers could be considered to be very simple, special-purpose robots, but what people really have in mind is something that will go around the house and collect the dirty dishes and clothes, sort and wash them, and then put them back to their proper places—after, of course, pressing and folding the clean clothes. All of these tasks are quite difficult, beyond the capabilities of the first few generations of robots.

Today, robots are not yet household objects. They show up in science fairs and factory floors, search-and-rescue missions, and other specialized events. But this will change. Sony has announced this to be the decade of the robot, and even if Sony is too optimistic, I do predict that robots will blossom forth during the first half of the twenty-first century.

FIGURE 6.2a and b

Home robots of the early twenty-first century.

Figure a, ER2, a prototype of a home robot from Evolution Robotics. Figure b, Sony's Aibo, a pet robot dog.

(Image of ER2 courtesy of Evolution Robotics. Image of "Three Aibos on a wall" courtesy of Sony Electronics Inc., Entertainment America, Robot Division.)

Robots will take many forms. I can imagine a family of robot appliances in the kitchen—refrigerator, pantry, coffeemaker, cooking, and dishwasher robots—all configured to communicate with one another and to transfer food, dishes, and utensils back and forth. The home servant robot wanders about, picking up dirty dishes, delivering them to the dishwasher robot. The dishwasher, in turn, delivers clean dishes and utensils to the robot pantry, which stores them until needed by person or robot. The pantry, refrigerator, and cooking robots work smoothly to prepare the day's menu and, finally, place the completed meal onto dishes provided by the pantry robot.

Some robots will take care of children by playing with them, reading to them, singing songs. Educational toys are already doing this, and the sophisticated robot could act as a powerful tutor, starting with the alphabet, reading, and arithmetic, but soon expanding to almost any topic. Neal Stephenson's science fiction novel, *The Diamond Age*, does a superb job of showing how an interactive book, *The Young Lady's Illustrated Primer,* can take over the entire education of young girls from age four through adulthood. The illustrated primer is still some time in the future, but more limited tutors are already in existence. In addition to education, some robots will do household chores: vacuuming, dusting, and cleaning up. Eventually their range of abilities will expand. Some may end up being built into homes or furniture. Some will be mobile, capable of wandering about on their own.

These developments will require a coevolutionary process of adaptation for both people and devices. This is common with our technologies: we reconfigure the way we live and work to make things possible for our machines to function. The most dramatic coevolution is the automobile system, for which we have altered our homes to include garages and driveways sized and equipped for the automobile, and built a massive worldwide highway system, traffic signaling systems, pedestrian passageways, and huge parking lots. Homes, too, have been transformed to accommodate the multiple wires and pipes of the ever-increasing infrastructure of modern life: hot and cold water, waste return, air vents to the roof, heating and cooling ducts, electricity, telephone, television, internet and home computer and entertainment networks. Doors have to be wide enough for our furniture, and many homes have to accommodate wheelchairs and people using walkers. Just as we have accommodated the home for all these changes, I expect modification to accommodate robots. Slow modification, to be sure, but as robots increase in usefulness, we will ensure their success by minimizing obstacles and, eventually, building charging stations, cleaning and maintenance places, and so on. After all, the vacuum cleaner robot will need a place to empty its dirt, and the garbage robot will need to be able to carry the garbage outside the home. I wouldn't be surprised to see robot quarters in homes, that is, specially built niches where the robots can reside, out of the way, when they are not active. We have closets and pantries for today's appliances, so why not ones especially equipped for robots, with doors that can be controlled by the robot, electrical outlets, interior lights so robots can see to clean themselves (and plug themselves into the outlets), and waste receptacles where appropriate.

Robots, especially at first, will probably require smooth floors, without obstacles. Door thresholds might have to be eliminated or minimized. Some locations—especially stairways—might have to be especially marked, perhaps with lights, infrared transmitters, or simply special reflective tape. Barcodes or distinctive markers posted here and there in the home would enormously simplify the robot's ability to recognize its location.

Consider how a servant robot might bring a drink to its owner. Ask for a can of soda, and off goes the robot, obediently making its way to the kitchen and the refrigerator, which is where the soda is kept. Understanding the command and navigating to the refrigerator are relatively simple. Figuring out how to open the door, find the can, and extract it is not so simple. Giving the servant robot the dexterity, the strength, and the non-slip wheels that would allow it to pull open the refrigerator door is quite a feat. Providing the vision system that can find the soda, especially if it is completely hidden behind other food items, is difficult, and then figuring out how to extract the can without destroying objects in the way is beyond today's capabilities in robot arms.

How much simpler it would be if there were a drink dispenser robot tailored to the needs of the servant robot. Imagine a drink-dispensing robot appliance capable of holding six or twelve cans, refrigerated, with an automatic door and a push-arm. The servant robot could go to the drink robot, announce its presence and its request (probably by an infrared or radio signal), and place its tray in front of the dispenser. The drink robot would slide open its door, push out a can, and close the door again: no complex vision, no dexterous arm, no forceful opening of the door. The servant robot would receive the can on its tray, and then go back to its owner.

In a similar way, we might modify the dishwasher to make it easier for a home robot to load it with dirty dishes, perhaps give it special trays with designated slots for different dishes. But as long as we are doing that, why not make the pantry a specialized robot, one capable of removing the clean dishes from the dishwasher and storing them for later use? The special trays would help the pantry as well. Perhaps the pantry could automatically deliver cups to the coffeemaker and plates to the home cooking robot, which is, of course, connected to refrigerator, sink, and trash. Does this sound far-fetched? Perhaps, but, in fact, our household appliances are already complex, many of them with multiple connections to services. The refrigerator has connections to electric power and water. Some are already connected to the internet. The dishwasher and clothes washer have electricity, water and sewer connections. Integrating these units so that they can work smoothly with one another does not seem all that difficult.

I imagine that the home will contain a number of specialized robots: the servant is perhaps the most general purpose, but it would work together with a cleaning robot, the drink dispensing robot, perhaps some outside gardening robots, and a family of kitchen robots, such as dishwasher, coffee-making, and pantry robots. As these robots are developed, we will probably also design specialized objects in the home that simplify the tasks for the robots, coevolving robot and home to work smoothly together. Note that the end result will be better for people as well. Thus, the drink dispenser robot would allow anyone to walk up to it and ask for a can, except that you wouldn't use infrared or radio, you might push a button or perhaps just ask.

I am not alone in imagining this coevolution of robots and homes. Rodney Brooks, one of the world's leading roboticists, head of the MIT Artificial Intelligence Laboratory and founder of a company that builds home and commercial robots, imagines a rich ecology of environments and robots, with specialized ones living on devices, each responsible to keep its domain clean: one does the bathtub, another the toilet; one does windows, another manipulates mirrors. Brooks even contemplates a robot dining room table, with storage area and dishwasher built into its base so that "when we want to set the table, small robotic arms, not unlike the ones in a jukebox, will bring the required dishes and cutlery out onto the place settings. As each course is finished, the table and its little robot arms would grab the plates and devour them into the large internal volume underneath."

What should a robot look like? Robots in the movies often look like people, with two legs, two arms, and a head. But why? Form should follow function. The fact that we have legs allows us to navigate irregular terrain, something an animal on wheels could not do. The fact that we have two hands allows us to lift and manipulate, with one hand helping the other. The humanoid shape has evolved over eons of interaction with the world to cope efficiently and effectively with it. So, where the demands upon a robot are similar to those upon people, having a similar shape might be sensible.

If robots don't have to move—such as drink, dishwasher, or pantry robots—they need not have any means of locomotion, neither legs nor wheels. If the robot is a coffeemaker, it should look like a coffeemaker, modified to allow it to connect to the dishwasher and pantry. Robot vacuum cleaners and lawn mowers already exist, and their appearance is perfectly suited to their tasks: small, squat devices, with wheels (see figure 6.3). A robot car should look like a car. It is only the general-purpose home servant robots that are apt to look like animals or humans. The robot dining room table envisioned by Brooks would be especially bizarre, with a large central column to house the dishes and dishwashing equipment (complete with electric power, water and sewer connections). The top of the table would have places for the robot arms to manipulate the dishes and probably some stalk to hold the cameras that let the arms know where to place and retrieve the dishes and cutlery.

Should a robot have legs? Not if it only has to maneuver about on smooth surfaces—wheels will do for this; but if it has to navigate irregular terrain or stairs, legs would be useful. In this case, we can expect the first legged robots to have four or six legs: balancing is far simpler for four- and six-legged creatures than for those with only two legs.

If the robot is to wander about a home and pick up after the occupants, it probably will look something like an animal or a person: a

FIGURE 6.3

What should a robot look like?

The Roomba is a vacuum cleaner, its shape appropriate for running around the floor and maneuvering itself under the furniture. This robot doesn't look like either a person or an animal, nor should it: its shape fits the task.

(Courtesy of iRobot Inc.)

body to hold the batteries and to support the legs, wheels, or tracks for locomotion; hands to pick up objects; and cameras (eyes) on top where they can better survey the environment. In other words, some robots will look like an animal or human, not because this is cute, but because it is the most effective configuration for the task. These robots will probably look something like R2D2 (figure 6.1): a cylindrical or rectangular body on top of some wheels, tracks, or legs; some form of manipulable arm or tray; and sensors all around to detect obstacles, stairs, people, pets, other robots, and, of course the objects they are supposed to interact with. Except for pure entertainment value, it is difficult to understand why we would ever want a robot that looked like C3PO.

In fact, making a robot humanlike might backfire, making it less acceptable. Masahiro Mori, a Japanese roboticist, has argued that we are least accepting of creatures that look very human, but that perform badly, a concept demonstrated in film and theater by the terrifying nature of zombies and monsters (think of Frankenstein's monster) that take on human form, but with inhuman movement and ghastly appearance. We are not nearly so dismayed—or frightened—by nonhuman shapes and forms. Even perfect replicas of humans might be problematic, for even if the robot could not be distinguished from humans, this very lack of distinction can lead to emotional angst (a theme explored in many a science fiction novel, especially Philip K. Dick's *Do Androids Dream of Electric Sheep?* and, in movie version, *Blade Runner*). According to this line of argument, C3PO gets away with its humanoid form because it is so clumsy, both in manner and behavior, that it appears more cute or even irritating than threatening.

Robots that serve human needs—for example, robots as pets—should probably look like living creatures, if only to tap into our visceral system, which is prewired to interpret human and animal body language and facial expressions. Thus, an animal or a childlike shape together with appropriate body actions, facial expressions, and sounds will be most effective if the robot is to interact successfully with people.

Affect and Emotion in Robots

What emotions will a robot need to have? The answer depends upon the sort of robot we are thinking about, the tasks it is to perform, the nature of the environment, and what its social life is like. Does it interact with other robots, animals, machines, or people? If so, it will need to express its own emotional state as well as to assess the emotions of the people and animals it interacts with.

Think about the average, everyday home robot. These don't yet exist, but some day the house will become populated with robots. Some home robots will be fixed in place, specialized, such as kitchen robots: for example, the pantry, dishwasher, drink dispenser, food dispenser, coffeemaker, or cooking unit robots. And, of course, clothes washer, drier, iron, and clothes-folding robots, perhaps coupled to wardrobe robots. Some will be mobile, but also specialized, such as the robots that vacuum the floors and mow the lawn. But probably we will also have at least one general-purpose robot: the home servant robot, that brings us coffee, cleans up, does simple errands, and looks after and supervises the other robots. It is the home robot that is of most interest, because it will have to be the most flexible and advanced.

Servant robots will need to interact with us and with the other robots of the house. For the other robots, they could use wireless communication. They could discuss the jobs they were doing, whether or not they were overloaded or idle. They could also state when they were running low on supplies and when they sensed difficulties, problems, or errors and call upon one another for help. But what about when robots interact with people? How will this happen?

Servant robots need to be able to communicate with their owners. Some way of issuing commands is needed, some way of clarifying the ambiguities, changing a command in midstream ("Forget the coffee, bring me a glass of water instead"), and dealing with all of the complexities of human language. Today, we can't do that, so robots that are built now will have to rely upon very simple commands or even some sort of remote controller, where a person pushes the appropriate buttons, generates a well-structured command, or selects actions from a menu. But the time will come when we can interact in speech, with the robots understanding not just the words but the meanings behind them.

When should a robot volunteer to help its owners? Here, robots will need to be able to assess the emotional state of people. Is someone struggling to do a task? The robot might want to volunteer to help. Are the people in the house arguing? The robot might wish to go to some other room, out of the way. Did something bring pleasure? The robot might wish to remember that, so it could do it again when appropriate. Was an action poorly done, so the person showed disappointment? Perhaps the action could be improved, so that next time the robot would produce better results. For all these reasons, and more, the robot will need to be designed with the ability to read the emotional state of its owners.

A robot will need to have eyes and ears (cameras and microphones) to read facial expressions, body language, and the emotional components of speech. It will have to be sensitive to tones of voice, the tempo of speech, and its amplitude, so that it can recognize anger, delight, frustration, or joy. It needs to be able to recognize scolding voices from praising ones. Note that all of these states can be recognized just by their sound quality without the need to recognize the words or language. Notice that you can determine other people's emotional states just by the tone of voice alone. Try it: Make believe you are in any one of those states—angry, happy, scolding, or praising—and express yourself while keeping your lips firmly sealed. You can do it entirely with the sounds, without speaking a word. These are universal sound patterns.

Similarly, the robot should display its emotional state, much as a person does (or, perhaps more appropriately, as a pet dog or child does), so that the people with whom it is interacting can tell when a request is understood, when it is something easy to do, difficult to do, or perhaps even when the robot judges it to be inappropriate. Similarly, the robot should show pleasure and displeasure, an energetic appearance or exhaustion, confidence or anxiety when appropriate. If it is stuck, unable to complete a task, it should show its frustration. It will be as valuable for the robot to display its emotional state as it is for people to do so. The expressions of the robot will allow us humans to understand the state of the robot, thereby learning which tasks are appropriate for it, which are not. As a result, we can clarify instructions or even offer help, eventually learning to take better advantage of the robot's capabilities.

Many people in the robotics and computer research community believe that the way to display emotions is to have a robot decide whether it is happy or sad, angry or upset, and then display the appropriate face, usually an exaggerated parody of a person in those states. I argue strongly against this approach. It is fake, and, moreover, it looks fake. This is not how people operate. We don't decide that we are happy, and then put on a happy face, at least not normally. This is what we do when we are trying to fool someone. But think about all those professionals who are forced to smile no matter what the circumstance: they fool no one—they look just like they are forcing a smile, as indeed they are.

The way humans show facial expression is by automatic innervation of the large number of muscles involved in controlling the face and body. Positive affect leads to relaxation of some muscle groups, automatic pulling up of many facial muscles (hence the smile, raised eyebrows and cheeks, etc.), and a tendency to open up and draw closer to the positive event or thing. Negative affect has the opposite impact, causing withdrawal, to push away. Some muscles are tensed, and some of the facial muscles pull downward (hence the frown). Most affective states are complex mixtures of positive and negative valence, at differing levels of arousal, with some residue of the immediately previous states. The resulting expressions are rich and informative. And real.

Fake emotions look fake: we are very good at detecting false attempts to manipulate us. Thus, many of the computer systems we interact with—the ones with cute, smiling helpers and artificially sweet voices and expressions—tend to be more irritating than useful. "How do I turn this off?" is a question often asked of me, and I have become adept at disabling them, both in my own computers or those of others who seek to be released from the irritation.

I have argued that machines should indeed both have and display emotions, the better for us to interact with them. This is precisely why the emotions need to appear as natural and ordinary as human emotions. They must be real, a direct reflection of the internal states and processing of a robot. We need to know when a robot is confident or confused, secure or worried, understanding our queries or not, working on our request or ignoring us. If the facial and body expressions reflect the underlying processing, then the emotional displays will seem genuine precisely because they are real. Then we can interpret their state, they can interpret ours, and the communication and interaction will flow ever more smoothly.

I am not the only person to have reached this conclusion. MIT Professor Rosalind Picard once said, talking about whether robots should have emotions, "I wasn't sure they had to have emotions until I was writing up a paper on how they would respond intelligently to our emotions without having their own. In the course of writing that paper, I realized it would be a heck of a lot easier if we just gave them emotions."

Once robots have emotions, then they need to be able to display them in a way that people can interpret—that is, as body language and facial expressions similar to human ones. Thus, the robot's face and body should have internal actuators that act and react like human muscles according to the internal states of the robot. People's faces are richly endowed with muscle groups in chin, lips, nostrils, eyebrows, forehead, cheeks, and so on. This complex of muscles makes for a sophisticated signaling system, and if robots were created in a similar way, the features of the face will naturally smile when things are going well and frown when difficulties arise. For this purpose, robot designers need to study and understand the complex workings of human expressions, with its very rich set of muscles and ligaments tightly intertwined with the affective system.

Displaying full facial emotions is actually very difficult. Figure 6.4 shows Leonardo, Professor Cynthia Breazeal's robot at the MIT Media Laboratory, designed to control a vast array of facial features, neck, body, and arm movements, all the better to interact socially and emotionally with us. There is a lot going on inside our bodies, and much the same complexity is required within the faces of robots.

But what of the underlying emotional states? What should these be? As I've discussed, at the least, the robot should be cautious of heights, wary of hot objects, and sensitive to situations that might lead to hurt or injury. Fear, anxiety, pain, and unhappiness might all be appropriate states for a robot. Similarly, it should have positive states,

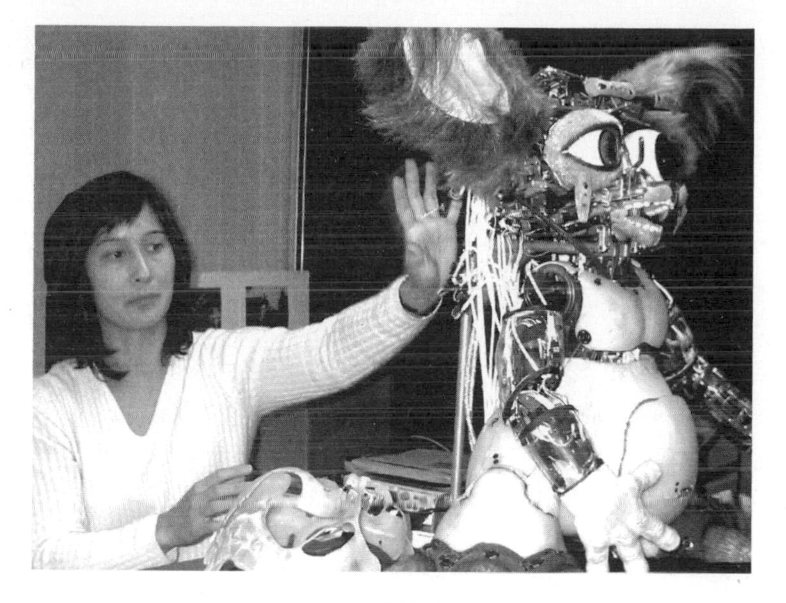

FIGURE 6.4

The complexity of robot facial musculature.

MIT Professor Cynthia Breazeal with her robot Leonardo.

(Photograph by author.)

including pleasure, satisfaction, gratitude, happiness and pride, which would enable it to learn from its actions, to repeat the positive ones and improve, where possible.

Surprise is probably essential. When what happens is not what is expected, the surprised robot should interpret this as a warning. If a room unexpectedly gets dark, or maybe the robot bumps into something it didn't expect, a prudent response is to stop all movement and figure out why. Surprise means that a situation is not as anticipated, and that planned or current behavior is probably no longer appropriate—hence, the need to stop and reassess.

Some states, such as fatigue, pain, or hunger, are simpler, for they do not require expectations or predictions, but rather simple monitoring of internal sensors. (Fatigue and hunger are technically not affective states, but they can be treated as if they were.) In the human, sensors of physical states signal fatigue, hunger, or pain. Actually, in people, pain is a surprisingly complex system, still not well understood. There are millions of pain receptors, plus a wide variety of brain centers involved in interpreting the signals, sometimes enhancing sensitivity, sometimes suppressing it. Pain serves as a valuable warning system, preventing us from damaging ourselves and, if we are injured, acting as a reminder not to stress the damaged parts further. Eventually it might be useful for robots to feel pain when motors or joints were strained. This would lead robots to limit their activities automatically, and thus protect themselves against further damage.

Frustration would be a useful affect, preventing servant robots from getting stuck doing a task to the neglect of its other duties. Here is how it would work. I ask the servant robot to bring me a cup of coffee. Off it goes to the kitchen, only to have the coffee robot explain that it can't give any because it lacks clean cups. Then the coffeemaker might ask the pantry robot for more cups, but suppose that it, too, didn't have any. The pantry would have to pass on the request to the dishwasher robot. And now suppose that the dishwasher didn't have any dirty ones it could wash. The dishwasher would ask the servant robot to search for dirty cups so that it could wash them, give them to the pantry, which would feed them to the coffeemaker, which in turn would give the coffee to the servant robot. Alas, the servant would have to decline the dishwasher's request to wander about the house: it is still busy at its main task—waiting for coffee.

This situation is called "deadlock." In this case, nothing can be done because each machine is waiting for the next, and the final machine is waiting for the first. This particular problem could be solved by giving the robots more and more intelligence, learning how to solve each new problem, but problems always arise faster than designers can anticipate them. These deadlock situations are difficult to eliminate because each one arises from a different set of circumstances. Frustration provides a general solution.

Frustration is a useful affect for both humans and machines, for when things reach that point, it is time to quit and do something else.

The servant robot should get frustrated waiting for the coffee, so it should temporarily give up. As soon as the servant robot gives up the quest for coffee, it is free to attend to the dishwasher's request, go off and find the dirty coffee cups. This would automatically solve the deadlock: the servant robot would find some dirty cups, deliver them to the dishwasher, which would eventually let the coffeemaker make the coffee and let me get my coffee, although with some delay.

Could the servant robot learn from this experience? It should add to its list of activities the periodic collection of dirty dishes, so that the dishwasher/pantry would never run out again. This is where some pride would come in handy. Without pride, the robot doesn't care: it has no incentive to learn to do things better. Ideally, the robot would take pride in avoiding difficulties, in never getting stuck at the same problem more than once. This attitude requires that robots have positive emotions, emotions that make them feel good about themselves, that cause them to get better and better at their jobs, to improve, perhaps even to volunteer to do new tasks, to learn new ways of doing things. Pride in doing a good job, in pleasing their owners.

Machines That Sense Emotion

> The extent to which emotional upsets can interfere with mental life is no news to teachers. Students who are anxious, angry, or depressed don't learn; people who are caught in these states do not take in information efficiently or deal with it well.
>
> —*Daniel Goleman,*
> Emotional Intelligence

Suppose machines could sense the emotions of people. What if they were as sensitive to the moods of their users as a good therapist might be? What if an electronic, computer-controlled educational system

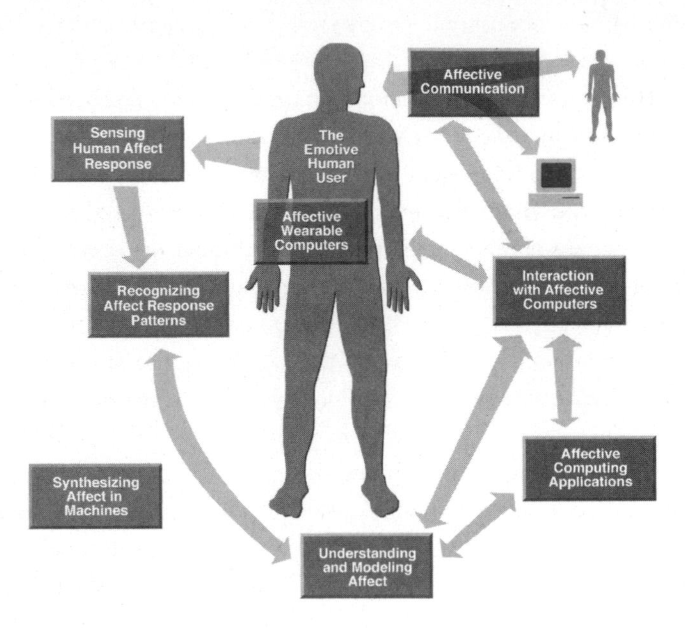

FIGURE 6.5

MIT's Affective Computing program.

The diagram indicates the complexity of the human affective system and the challenges required to monitor affect properly. From the work of Prof. Rosalind Picard of MIT.

(Drawing courtesy of Roz Picard and Jonathan Klein.)

could sense when the learner was doing well, was frustrated, or was proceeding appropriately? Or what if the home appliances and robots of the future could change their operations according to the moods of their owners? What then?

Professor Rosalind Picard at the MIT Media Laboratory leads a research effort entitled "Affective Computing," an attempt to develop machines that can sense the emotions of the people with whom they are interacting, and then respond accordingly. Her research group has made considerable progress in developing measuring devices to sense fear and anxiety, unhappiness and distress. And, of course, satisfaction and happiness. Figure 6.5 is taken from their web site and demonstrates the variety of issues that must be addressed.

How are someone's emotions sensed? The body displays its emotional state in a variety of ways. There are, of course, facial expressions and body language. Can people control their expressions? Well, yes, but the visceral layer works automatically, and although the behavioral and reflective levels can try to inhibit visceral reaction, complete suppression does not appear to be possible. Even the most controlled person, the so-called poker-face who keeps a neutral display of emotional responses no matter what the situation, still has micro-expressions—short, fleeting expressions that can be detected by trained observers.

In addition to the responses of one's musculature, there are many physiological responses. For example, although the size of the eye's pupil is affected by light intensity, it is also an indicator of emotional arousal. Become interested or emotionally aroused, and the pupil widens. Work hard on a problem, and it widens. These responses are involuntary, so it is difficult—probably impossible—for a person to control them. One reason professional gamblers sometimes wear tinted eyeglasses even in dark rooms is to prevent their opponents from detecting changes in the size of their pupils.

Heart rate, blood pressure, breathing rate, and sweating are common measures that are used to derive affective state. Even amounts of sweating so small that the person can be unaware of it can trigger a change in the skin's electrical conductivity. All of these measures can readily be detected by the appropriate electronics.

The problem is that these simple physiological measures are indirect measures of affect. Each is affected by numerous things, not just by affect or emotion. As a result, although these measures are used in many clinical and applied settings, they must be interpreted with care. Thus, consider the workings of the so-called lie detector. A lie detector is, if anything, an emotion detector. The method is technically called "polygraph testing" because it works by simultaneously recording and graphing multiple physiological measures such as heart rate, breathing rate, and skin conductance. A lie detector does not detect falsehoods; it detects a person's affective response to a series of questions being asked by the examiner, where some of the answers are assumed to be truthful (and thus show low affective responses) and some deceitful (and thus show high affective arousal). It is easy to see why lie detectors are so controversial. Innocent people might have large emotional responses to critical questions while guilty people might show no response to the same questions.

Skilled operators of lie detectors try to compensate for these difficulties by the use of control questions to calibrate a person's responses. For example, by asking a question to which they expect a lie in response, but that is not relevant to the issue at hand, they can see what a lie response looks like in the person being tested. This is done by interviewing the suspect and then developing a series of questions designed to ferret out normal deviant behavior, behavior in which the examiner has no interest, but where the suspect is likely to lie. One question commonly used in the United States is "Did you ever steal something when you were a teenager?"

Because lie detectors record underlying physiological states associated with emotions rather than with lies, they are not very reliable, yielding both misses (when a lie is not detected because it produces no emotional response) and false alarms (when the nervous suspect produces emotional responses even though he or she is not guilty). Skilled operators of these machines are aware of the pitfalls, and some use the lie detector test as a means of eliciting a confession: people who truly believe the lie detector can "read minds" might confess just because of their fear of the test. I have spoken to skilled operators who readily agree to the critique I just provided, but are proud of their record of eliciting voluntary confessions. But even innocent people have sometimes confessed to crimes they did not commit, strange as this might seem. The record of accuracy is flawed enough that the National Research Council of the United States National Academies performed a lengthy, thorough study and concluded that polygraph testing is too flawed for security screening and legal use.

SUPPOSE WE could detect a person's emotional state, then what? How should we respond? This is a major, unsolved problem. Consider the classroom situation. If a student is frustrated, should we try to remove the frustration, or is the frustration a necessary part of learning? If an automobile driver is tense and stressed, what is the appropriate response?

The proper response to an emotion clearly depends upon the situation. If a student is frustrated because the information provided is not clear or intelligible, then knowing about the frustration is important to the instructor, who presumably can correct the problem through further explanation. (In my experience, however, this often fails, because an instructor who causes such frustration in the first place is usually poorly equipped to understand how to remedy the problem.)

If the frustration is due to the complexity of the problem, then the proper response of a teacher might be to do nothing. It is normal and proper for students to become frustrated when attempting to solve problems slightly beyond their ability, or to do something that has never been done before. In fact, if students aren't occasionally frustrated, it probably is a bad thing—it means they aren't taking enough risks, they aren't pushing themselves sufficiently.

Still, it probably is good to reassure frustrated students, to explain that some amount of frustration is appropriate and even necessary. This is a good kind of frustration that leads to improvement and learning. If it goes on too long, however, the frustration can lead students to give up, to decide that the problem is above their ability. Here is where it is necessary to offer advice, tutorial explanations, or other guidance.

What of frustrations shown by students that have nothing to do with the class, that might be the result of some personal experience, outside the classroom? Here it isn't clear what to do. The instructor, whether person or machine, is not apt to be a good therapist. Expressing sympathy might or might not be the best or most appropriate response.

Machines that can sense emotions are an emerging new frontier of research, one that raises as many questions as it addresses, both in how machines might detect emotions and in how to determine the most appropriate way of responding. Note that while we struggle to determine how to make machines respond appropriately to signs of emotions, people aren't particularly good at it either. Many people have great difficulty responding appropriately to others who are experiencing emotional distress: sometimes their attempts to be helpful make the problem worse. And many are surprisingly insensitive to the emotional states of others, even people whom they know well. It is natural for people under emotional strain to try to hide the fact, and most people are not experts in detecting emotional signs.

Still, this is an important research area. Even if we are never able to develop machines that can respond completely appropriately, the research should inform us both about human emotion and also about human-machine interaction.

Machines That Induce Emotion in People

It is surprisingly easy to get people to have an intense emotional experience with even the simplest of computer systems. Perhaps the earliest such experience was with Eliza, a computer program developed by the MIT computer scientist Joseph Weizenbaum. Eliza was a simple program that worked by following a small number of conversational scripts that had been prepared in advance by the programmer (originally, this was Weizenbaum). By following these scripts, Eliza could interact with a person on whatever subject the script had prepared it for. Here is an example. When you started the program, it would greet you by saying: "Hello. I am ELIZA. How can I help you?" If you responded by typing: "I am concerned about the increasing level of violence in the world," Eliza would respond: "How long have you been concerned about the increasing level of violence in the world?" That's a relevant question, so a natural reply would be something like, "Just the last few months," to which Eliza would respond, "Please go on."

You can see how you might get captured by the conversation: your concerns received sympathetic responses. But Eliza has no understanding of language. It simply finds patterns and responds appropriately (saying "Please go on" when it doesn't recognize the pattern). Thus, it is easy to fool Eliza by typing: "I am concerned about abc, def, and for that matter, ghi," to which Eliza would dutifully reply: "How long have you been concerned about abc, def, and for that matter, ghi?"

Eliza simply recognizes the phrase "I am concerned about X" and replies, "How long have you been concerned about X?" with absolutely no understanding of the words.

Because most people took Eliza seriously, they did not try to trick it. Instead, they took each reply seriously, pondering their hidden meanings. As a result, they would sometimes discover themselves in deep, philosophical debate or, in the most popular version of Eliza that was scripted to act like a psychotherapist, they would find themselves discussing their most intimate secrets.

In the mid-1960s, one of my long-term friends and research collaborators, Daniel Bobrow, was a research scientist at BBN, a Cambridge, Massachusetts company doing research in artificial intelligence, computer networks, and acoustics. Bobrow told me that a programmer, having worked hard to get Eliza working on a BBN computer, finally finished about 5:00 A.M. Saturday morning and then went home, leaving the program running on the computer. (In those days—this was 1966—personal computers were unheard of, and only advanced companies, such as BBN, even had computers for its employees to use, which is one reason programmers had to work at night to get computer time.)

The vice president of Bobrow's division came in later that Saturday morning to demonstrate the computer to some customers. Seeing that the computer was running, he assumed that Bobrow was using it from home, and, according to Bobrow, "decided to use the 'obvious' communication link I must have left going" to chat with him. So the VP typed some questions for Bobrow. But Bobrow was at home, asleep, and what ensued was actually taking place between Eliza and the VP, who assumed these responses came from Bobrow:

VP: Do you think that I can use the computer this morning?
Eliza: Why do you ask?
VP: I might be able to make some additional Telcom sales.
Eliza: Why aren't you so sure?
VP: My prospective customers have never seen a demonstration of the system.
Eliza: Does this mean very much to you?
VP: Of course it does.
Eliza: You seem very positive about that.
VP: This is ridiculous.
Eliza: What do you mean, this is ridiculous?
VP: Is there any reason why I can't use the computer now?
Eliza: Tell me more about your prospective customers.
VP: Please dial me up on 491-1850

In those days, computers were not very powerful, and the only way that the Eliza program could tell when a person had finished a sentence was by noticing a period. Up to that last line, the VP had finished each line with a period, but this time he neglected it. As a result, the computer kept waiting for more typing and, therefore, didn't answer. According to Bobrow, "This so infuriated the VP, who thought I was playing games with him, that he called me up, woke me from a deep sleep, and said: 'Why are you being so snotty to me?' 'What do you mean I am being snotty to you?,' I answered." Then, Bobrow told me, "The VP angrily read me the dialog that 'we' had been having, and couldn't get any response but laughter from me. It took a while to convince him it really was a computer."

As Bobrow told me when I discussed this interaction with him, "You can see he cared a lot about the answers to his questions, and what he thought were my remarks had an emotional effect on him." We are extremely trusting, which makes us very easy to fool, and very angry when we think we aren't being taken seriously.

The reason Eliza had such a powerful impact is related to the discussions in chapter 5 on the human tendency to believe that any intelligent-seeming interaction must be due to a human or, at least, an intelligent presence: anthropomorphism. Moreover, because we are trusting, we tend to take these interactions seriously. Eliza was written a long time ago, but its creator, Joseph Weizenbaum, was horrified by the seriousness with which his simple system was taken by so many people who interacted with it. His concerns led him to write *Computer Power and Human Reason,* in which he argued most cogently that these shallow interactions were detrimental to human society.

We have come a long way since Eliza was written. Computers of today are thousands of times more powerful than they were in the 1960s and, more importantly, our knowledge of human behavior and psychology has improved dramatically. As a result, today we can write programs and build machines that, unlike Eliza, have some true understanding and can exhibit true emotions. However, this doesn't mean that we have escaped from Weizenbaum's concerns. Consider Kismet.

Kismet, whose photograph is shown in figure 6.6, was developed by a team of researchers at the MIT Artificial Intelligence Laboratory and reported upon in detail in Cynthia Breazeal's *Designing Sociable Robots.*

Recall that the underlying emotions of speech can be detected without any language understanding. Angry, scolding, pleading, consoling, grateful, and praising voices all have distinctive pitch and loudness contours. We can tell which of these states someone is in even if they are speaking in a foreign language. Our pets can often detect our moods through both our body language and the emotional patterns within our voices.

Kismet uses these cues to detect the emotional state of the person

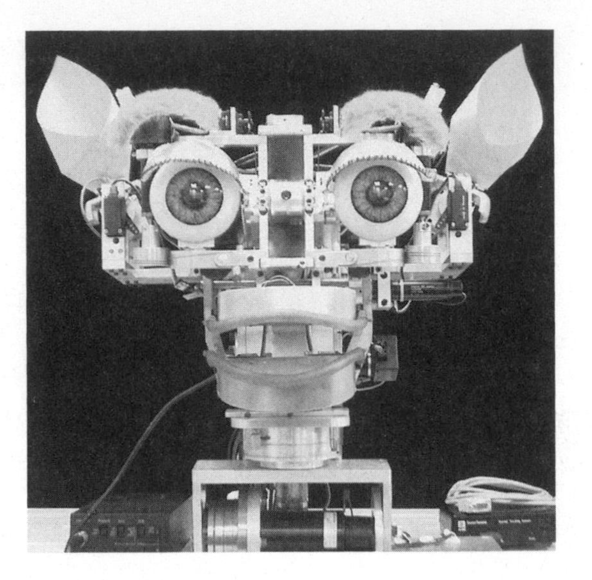

FIGURE 6.6

Kismet, a robot designed for social interactions, looking surprised.

(Image courtesy of Cynthia Breazeal.)

with whom it is interacting. Kismet has video cameras for eyes and a microphone with which to listen. Kismet has a sophisticated structure for interpreting, evaluating, and responding to the world—shown in figure 6.7—that combines perception, emotion, and attention to control behavior. Walk up to Kismet, and it turns to face you, looking you straight in the eyes. But if you just stand there and do nothing else, Kismet gets bored and looks around. If you do speak, it is sensitive to the emotional tone of the voice, reacting with interest and pleasure to encouraging, rewarding praise and with shame and sorrow to scolding. Kismet's emotional space is quite rich, and it can move its head, neck, eyes, ears, and mouth to express emotions. Make it sad, and its ears droop. Make it excited and it perks up. When unhappy, the head droops, ears sag, mouth turns down.

Interacting with Kismet is a rich, engaging experience. It is difficult to believe that Kismet is all emotion, with no understanding. But walk up to it, speak excitedly, show it your brand-new watch, and Kismet

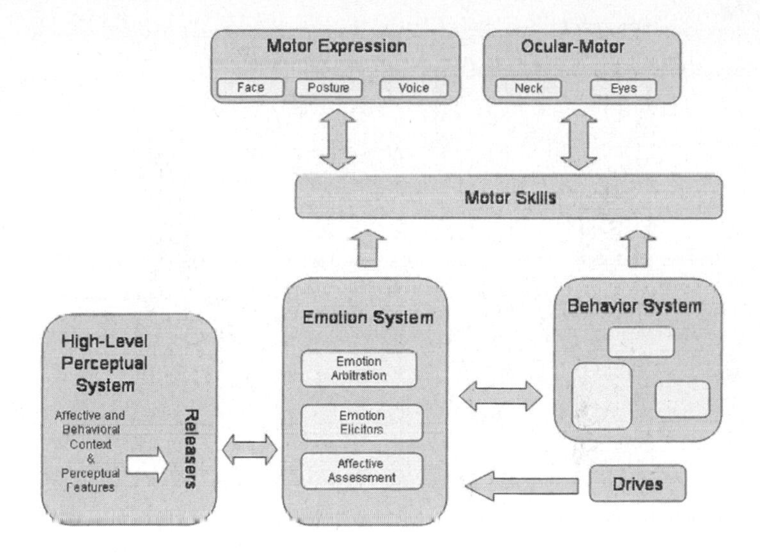

FIGURE 6.7

Kismet's emotional system.

The heart of Kismet's operation is in the interaction of perception, emotion, and behavior.

(Figure redrawn, slightly modified with permission of Cynthia Breazeal, from http://www.ai.mit.edu/projects/sociable/emotions.html.)

responds appropriately: it looks at your face, then at the watch, then back at your face again, all the time showing interest by raising its eyelids and ears, and exhibiting perky, lively behavior. Just the interested responses you want from your conversational partner, even though Kismet has absolutely no understanding of language or, for that matter, your watch. How does it know to look at the watch? It doesn't, but it responds to movement so it looks at your rising hand. When the motion stops, it gets bored, and returns to look at your eyes. It shows excitement because it detected the tone of your voice.

Note that Kismet shares some characteristics of Eliza. Thus, although this is a complex system, with a body (well, a head and neck), multiple motors that serve as muscles, and a complex underlying model of attention and emotion, it still lacks any true understanding. Therefore, the interest and boredom that it shows toward people are simply programmed responses to changes—or the lack thereof—in the environment and responses to movement and physical aspects of speech. Although Kismet can sometimes keep people entranced for long periods, the enhancement is somewhat akin to that of Eliza: most of the sophistication is in the observer's interpretations.

Aibo, the Sony robot dog, has a far less sophisticated emotional repertoire and intelligence than Kismet. Nonetheless, Aibo has also proven to be incredibly engaging to its owners. Many owners of the robot dog band together to form clubs: some own several robots. They trade stories about how they have trained Aibo to do various tricks. They share ideas and techniques. Some firmly believe that their personal Aibo recognizes them and obeys commands even though it is not capable of these deeds.

When machines display emotions, they provide a rich and satisfying interaction with people, even though most of the richness and satisfaction, most of the interpretation and understanding, comes from within the head of the person, not from the artificial system. Sherry Turkle, both an MIT professor and a psychoanalyst, has summarized these interactions by pointing out, "It tells you more about us as human beings than it does the robots." Anthropomorphism again: we read emotions and intentions into all sorts of things. "These things push on our buttons whether or not they have consciousness or intelligence," Turkle said. "They push on our buttons to recognize them as though they do. We are programmed to respond in a caring way to these new kinds of creatures. The key is these objects want you to nurture them and they thrive when you pay attention."

Chapter Seven

The Future of Robots

SCIENCE FICTION CAN BE a useful source of ideas and information, for it is, in essence, detailed scenario development. Writers who have used robots in their stories have had to imagine in considerable detail just how they would function within everyday work and activities. Isaac Asimov was one of the earliest thinkers to explore the implications of robots as autonomous, intelligent creatures, equal (or superior) in intelligence and abilities to their human masters. Asimov wrote a sequence of novels analyzing the difficulties that would arise if autonomous robots populated the earth. He realized that a robot might inadvertently harm itself or others, both through its actions or, at times, through its lack of action. He therefore developed a set of postulates that might prevent these problems; but, as he did so, he also realized that they were often in conflict with one another. Some conflicts were simple: given a choice between preventing harm to itself or to a human, the robot should protect the human. But other conflicts were much more subtle, much more difficult. Eventually, he postulated three laws of robotics (laws one, two, and three) and wrote a sequence of stories to illustrate the dilemmas that robots would find themselves in, and how the three laws would allow them to handle these situations. These three laws dealt with the interaction of robots and people, but as his story line progressed into more complex situations, Asimov felt compelled to add an even more fundamental law dealing with the robots' relationship to humanity itself. This one was so fundamental that it had to come first; but, because he already had a law labeled First, this fourth law had to be labeled Zeroth.

Asimov's vision of people and of the workings of industry was strangely crude. It was only his robots that behaved well. When I reread his books in preparation for this chapter, I was surprised at the discrepancy between my fond memories of the stories and my response to them now. His people are rude, sexist, and naïve. They seem unable to converse unless they are insulting each other, fighting, or jeering. His fictional company, the *U.S. Robots and Mechanical Men Corporation* doesn't fare well either. It is secretive, manipulative, and allows no tolerance for error: make one mistake and the company would fire you. Asimov spent his entire life in a university. Maybe that is why he had such a weird view of the real world.

Nonetheless, his analysis of the reaction of society to robots—and of robots to humans—was interesting. He thought society would turn against robots; and, indeed, he wrote that "most of the world governments banned robot use on earth for any purpose other than scientific research between 2003 and 2007." (Robots, however, were allowed for space exploration and mining; and in Asimov's stories, these activities are widely deployed in the early 2000s, which allow the robot industry to survive and grow.) The Laws of Robotics are intended to reassure humanity that robots will not be a threat and will, moreover, always be subservient to humans.

Today, even our most powerful and functional robots are far from the stage of Asimov's. They do not operate for long periods without human control and assistance. Even so, the laws are an excellent tool for examining just how robots and humans should interact.

Asimov's Four Laws of Robotics

Zeroth Law: A robot may not injure humanity, or, through inaction, allow humanity to come to harm.

First law: A robot may not injure a human being, or, through inaction, allow a human being to come to harm, unless this would violate the Zeroth Law of Robotics.

Second Law: A robot must obey orders given it by human beings, except where such orders would conflict with the Zeroth or First Law.

Third Law: A robot must protect its own existence as long as such protection does not conflict with the Zeroth, First, or Second Law.

Many machines already have key aspects of the laws hard-wired into them. Let's examine how these laws are implemented.

The Zeroth Law—that "a robot may not injure humanity, or, through inaction, allow humanity to come to harm," is beyond current capability, for much the same reasons that Asimov did not need this law in his early stories: to determine just when an action—or lack of action—will harm all humanity is truly sophisticated, probably beyond the abilities of most people.

The first law—that "a robot may not injure a human being, or, through inaction, allow a human being to come to harm, unless this would violate the Zeroth Law of Robotics," could be labeled "safety." It isn't legal, let alone proper, to produce things that can hurt people. As a result, all machines today are designed with multiple safeguards to minimize the likelihood that they can harm by their action. Liability laws guarantee that robots—and machines in general—are outfitted with numerous safeguards to prevent their actions from harming people. Industrial and home robots have proximity and collision sensors. Even simple machines such as elevators and garage doors have sensors that stop them from closing on people. Today's robots try to avoid bumping into people or objects. Lawn mower and vacuum cleaner robots have sensing mechanisms that cause them to stop or back away whenever they bump into anything or come too close to an edge, such as a stairway. Industrial robots are often fenced off, so that people can't get near them when they are operating. Some have people detectors, so they stop when they detect someone nearby. Home robots have many mechanisms to minimize the chance of damage; but at the moment, most of them are so underpowered that they couldn't hurt even if they tried to. Moreover, the lawyers are very careful to guard against potential damage. One company sells a home robot that can be used to teach children by reading books to them and that can also serve as a home sentinel, wandering about the house, taking photographs of unexpected encounters and notifying its owners, by email if necessary (through its wireless internet connection, attaching the photographs along with the message, of course). Despite these intended applications, the robot comes with stern instructions that it is not to be used near children, nor is it to be left unattended in the house.

A lot of effort has gone into implementation of the safety provision of the first law. Most of this work can be thought of as applying to the visceral level, where fairly simple mechanisms are used to shut down the system if safety regulations are violated.

The second part of the law—do not allow harm through inaction—is quite difficult to implement. If determining how a machine's actions might affect people is difficult, trying to determine how the lack of an action might have an impact is even more so. This would be a reflective level implementation, for the robot would have to do considerable analysis and planning to determine when lack of action would lead to harm. This is beyond most capabilities today.

Despite the difficulties, some simple solutions to the problem do exist. Many computers are plugged into "non-interruptible power supplies" to avoid loss of data in cases of power failure. If the power failed and no action were taken, harm would occur, but in these cases, when the power fails, the power supply springs into action, switching to batteries, converting the battery voltage to the form the computer requires. It can also be set to notify people and to turn off the computer gracefully. Other safety systems are designed to act when normal processes have failed. Some automobiles have internal sensors that watch over the path of the car, adjusting engine power and braking to ensure that the auto keeps going as intended. Automatic speed control mechanisms attempt to keep a safe distance from the car in front, and lane-changing detectors are under investigation. All of these devices safeguard car and passengers when inaction would lead to accident.

Today, these devices are simple and the mechanisms built in. Still, one can see the beginnings of solutions to the inaction clause of the first law, even in these simple devices.

The second law—that "a robot must obey orders given it by human beings, except where such orders would conflict with the Zeroth or First Law," is about obeying people, in contrast to the first, which is about protecting them. In many ways, this law is trivial to implement, but for elementary reasons. Machines today do not have an independent mind, so they must obey orders: they have no choice but to follow the commands given them. If they fail, they face the ultimate punishment: they are shut off and sent to the repair shop.

Can a machine disobey the second law in order to protect the first law? Yes, but not with much subtlety. Command an elevator to take you to your floor, and it will refuse if it senses that a person or object is blocking the door. This, however, is the most trivial of ways to implement the law, and it fails when the situation has any sophistication. Actually, in cases where safety systems prevent a machine from following orders, usually a person can override the safety system to permit the operation to take place anyway. This has been the cause of many an accident in trains, airplanes, and factories. Maybe Asimov was correct: we should leave some decisions up to the machines.

Some automatically deployed safety systems are an example of the "through inaction" clause of the law. Thus, if the driver of an automobile steps on the brakes rapidly, but only depresses the brake pedal halfway, most autos would only slow halfway. The Mercedes Benz, however, considers this "harm through inaction," so when it detects a rapid brake deployment, it puts the brakes on full, assuming that the owner really wants to stop as soon as possible. This is a combination of the first and second laws: the first law, because it prevents harm to the driver; and the second law because it is violating the "instructions" to apply the brakes at half strength. Of course, this may not really be a violation of the instructions: the robot assumes that full power was intended, even if not commanded. Perhaps the robot is invoking a new rule: "Do what I mean, not what I say," an old concept from some early artificial intelligence computer systems.

Although the automatic application of brakes in an automobile is a partial implementation of the second law, the correct implementation would have the auto examine the roadway ahead and decide for itself just how much speed, braking, or steering ought to be applied. Once that happens, we will indeed have a full first and second law implementation. Once again, this is starting to happen. Some cars automatically slow up if they're too close to the car in front, even if the driver has not acted to slow the vehicle.

We don't yet have the case of conflicting orders, but soon we will have interacting robots, where the requests of one robot might conflict with the requests of the human supervisors. Then, determining precedence and priority will become important.

Once again, these are easy cases. Asimov had in mind situations where a car would refuse to drive: "I'm sorry, but the road conditions are too dangerous tonight." We haven't yet reached that point—but we will. Asimov's second law will be useful.

Least important of all the laws, so Asimov thought, was self-preservation—"a robot must protect its own existence as long as such protection does not conflict with the Zeroth, First, or Second Law"—so it is numbered three, last in the series. Of course, given the limited capability of today's machines, where laws one and two seldom apply, this law is of most importance today, for we would be most annoyed if our expensive robot damaged or destroyed itself. As a result, this law is easy to find in action within many existing machines. Remember those sensors that are built into robot vacuum cleaners to prevent them from falling down stairs? Also how they—and robot lawn mowers—have bump and obstacle detectors to avoid damage from collisions? In addition, many robots monitor their energy state and either go into "sleep" mode or return to a charging station when their energy level drops. Resolution of conflicts with the other laws is not well handled, except by the presence of human operators who are able to override safety parameters when circumstances warrant.

Asimov's Laws cannot be fully implemented until machines have a powerful and effective capability for reflection, including meta-knowledge (knowledge of its own knowledge) and self-awareness of its state, activities, and intentions. These raise deep issues of philosophy and science as well as complex implementation problems for engineers and programmers. Progress in this area is happening, but slowly.

Even with today's rather primitive devices, having some of the capabilities would be useful. Thus, in cases of conflict, there would be sensible overriding of the commands. Automatic controls in airplanes would look ahead to determine the implications of the path they are following so that they would change if it would lead to danger. Some planes have indeed flown into mountains while on automatic control, so the capability would have saved lives. In actuality, many automated systems already are beginning to do this kind of checking.

Even today's toy pet robots have some self-awareness. Consider a robot whose operation is controlled both by its "desire" to play with its human owner, but also to make sure that it doesn't exhaust its battery power. When low on energy, it will therefore return to its charging station, even if the human wishes to continue playing with it.

The greatest hurdles to our ability to implement something akin to Asimov's Laws are his underlying assumptions of autonomous operation and central control mechanisms that may not apply in today's systems.

Asimov's robots worked as individuals. Give a robot a task to do, and off it would go. In the few cases where he had robots work as a group, one robot was always in charge. Moreover, he never had people and robots working together as a team. We are more likely to want cooperative robots, systems in which people and robots or teams of robots work together, much as a group of human workers can work together at a task. Cooperative behavior requires a different set of assumptions than Asimov had. Thus, cooperative robots need rules that provide for full communication of intentions, current state, and progress.

Asimov's main failure, however, was his assumption that someone had to be in control. When he wrote his novels, it was common to assume that intelligence required a centralized coordinating and control mechanism with a hierarchical organizational structure beneath it. This is how armies have been organized for thousands of years: armies, governments, corporations, and other organizations. It was natural to assume that the same principle applied to all intelligent systems. But this is not the way of nature. Many natural systems, from the actions of ants and bees, to the flocking of birds, and even the growth of cities and the structure of the stock market, occur as a natural result of the interaction of multiple bodies, not through some central, coordinated control structure. Modern control theory has moved away from this assumption of a central command post. Distributed control is the hallmark of today's systems. Asimov assumed a central decision structure for each robot that decided how to act, guided by his laws. In fact, that is probably not how it will work: the laws will be part of the robot's architecture, distributed throughout the many modules of its mechanisms; lawful behavior will emerge from the interactions of the multiple modules. This is a modern concept, not understood while Asimov was writing, so it is no wonder he missed this development in our understanding of complex systems.

Still, Asimov was ahead of his time, thinking far ahead to the future. His stories were written in the 1940s and '50s, but in his novel *I, Robot*, he quotes the three laws of robotics from the 2058 edition of the *Handbook of Robotics*; thus, he looked ahead more than 100 years. By 2058, we may indeed need his laws. Moreover, as the analyses indicate, the laws are indeed relevant, and many systems today follow them, even if inadvertently. The difficult aspects have to do with damage due to lack of action, as well as with properly assessing the relative importance of following orders versus damage or harm to oneself, others, or humanity.

As machines become more capable, as they take over more and more human activities, working autonomously, without direct supervision, they will get entangled in the legal system, which will try to determine fault when accidents arise. Before this happens, it would be useful to have some sort of ethical procedure in place. There already are some safety regulations that apply to robots, but they are very primitive. We will need more.

It is not too early to think about the future difficulties that intelligent and emotional machines may give rise to. There are numerous practical, moral, legal, and ethical issues to think about. Most are still far in the future, but that is a good reason to start now—so that when problems arrive, we will be ready.

The Future of Emotional Machines and Robots: Implications and Ethical Issues

The development of smart machines that will take over some tasks now done by people has important ethical and moral implications. This point becomes especially critical when we talk about humanoid robots that have emotions and to which people might form strong emotional attachments.

What is the role of emotional robots? How will they interact with us? Do we really want machines that are autonomous, self-directed, with a wide range of behavior, a powerful intelligence, and affect and emotion? I think we do, for they can provide many benefits. Obviously, as with all technologies, there are dangers as well. We need to ensure that people always maintain oversight and control, that they serve human needs appropriately.

Will robot teachers replace human teachers? No, but they can complement them. Moreover, they could be sufficient in situations where there is no alternative—to enable learning while traveling, or while in remote locations, or when one wishes to study a topic for which there is not easy access to teachers. Robot teachers will help make lifelong learning a practicality. They can make it possible to learn no matter where one is in the world, no matter the time of day. Learning should take place when it is needed, when the learner is interested, not according to some arbitrary, fixed school schedule.

Many are bothered by these possibilities, so much so that they reject them out of hand as unethical, immoral. Although I do not do so, I do sympathize with their concerns. However, I see the development of intelligent machines as both inevitable and beneficial. Where will there be benefits? In such areas as doing dangerous tasks, driving automobiles, piloting commercial vessels, in education, in medicine, and in taking over routine work. Where might there be moral and ethical concerns? Pretty much in the same list of activities. Let me explore the beneficial aspects in more detail.

Consider some of the benefits. Robots could be—and to some extent already are—used in dangerous tasks, where people's lives are at risk. This includes such things as search-and-rescue operations, exploration, and mining. What are the problems? The major ones are likely to come from the use of robots to enhance illegal or unethical activities: robbery, murder, and terrorism.

Will robot cars replace the need for human drivers? I hope so. Every year, tens of thousands of people are killed, and hundreds of thousands seriously injured through motor vehicle accidents. Wouldn't it be nice if automobiles were as safe as commercial aviation? Here is where automated vehicles can be a wonderful saving. Moreover, automated vehicles could drive more closely to one another, helping to reduce traffic congestion, and they could drive more efficiently, helping to solve some of the energy issues associated with driving.

Driving an automobile is deceptively simple: most of the time it takes little skill. As a result, many are lulled into a false sense of security and self-confidence. But when danger arises, it does so rapidly, and then the distracted, the semiskilled, the untrained, and those temporarily impaired by drugs, alcohol, illness, fatigue, or sleep deprivation are often incapable of reacting properly in time. Even well-trained commercial drivers have accidents: automated vehicles will not reduce all accidents and injuries, but they stand a good chance of dramatically reducing the present toll. Yes, some people truly enjoy the sport of driving, but these could be accommodated on special roads, recreational areas, and race tracks. Automation of everyday driving would lead to loss of jobs for drivers of commercial vehicles, but with a saving of life, overall.

Robot tutors have great potential for changing the way we teach. Today's model is far too often that of a pedant lecturing at the front of the classroom, forcing students to listen to material they have no interest in, that appears irrelevant to their daily lives. Lectures and textbooks are the easiest way to teach from the point of view of the teacher, but the least effective for the learner. The most powerful learning takes place when well-motivated students get excited by a topic and then struggle with the concepts, learning how to apply them to issues they care about. Yes, struggle: learning is an active, dynamic process, and struggle is a part of it. But when students care about something the struggle is enjoyable. This is how great teaching has always taken place—not through lecturing, but through apprenticeship, coaching, and mentoring. This is how athletes learn. This is the essence of the attraction of video games, except that in games, what students learn is of little practical value. These methods are well known in the learning sciences, where they are called problem-based, inquiry-learning, or constructivist.

Here is where emotion plays its part. Students learn best when motivated, when they care. They need to be emotionally involved, to be drawn to the excitement of the topic. This is why examples, diagrams and illustrations, videos and animated illustrations are so powerful. Learning need not be a dull and dreary exercise, not even learning about what are normally considered dull and dreary topics: every topic can be made exciting, every topic excites the emotions of someone, so why not excite everyone? It is time for lessons to become alive, for history to be seen as a human struggle, for students to understand and appreciate the structure of art, music, science, and mathematics. How can these topics be made exciting? By making them relevant to the lives of each individual student. This is often most effective by having students put their skills to immediate application. Developing exciting, emotionally engaging, and intellectually effective learning experiences is truly a design challenge worthy of the best talent in the world.

Robots, machines, or computers can be of great assistance in instruction by providing the framework for motivated, problem-based learning. Computer learning systems can provide simulated worlds in which students can explore problems in science, literature, history, or the arts. Robot teachers can make it easy to search the world's libraries and knowledge bases. Human teachers will no longer have to lecture, but instead can spend their time as coaches and mentors, helping to teach not only the topic, but also how best to learn, so that the students will maintain their curiosity through life, as well as the ability to teach themselves when necessary. Human teachers are still essential, but they can play a different, much more supportive and constructive role than they do today.

Moreover, although I believe strongly that we could develop efficient robot tutors, perhaps as effective as Stephenson's *The Young Lady's Illustrated Primer* (see page 171), we would not have to abandon human teachers: automated tutors—whether books, machines, or robots—should act as supplements to human instruction. Even Stephenson writes in his novel that his star pupil knew nothing of the real world and of real people because she had spent far too much time locked up in the fantasy world of the Primer.

Robots in medicine? Yes, they could be used in all its aspects. In medicine, however, as in many other activities, I foresee this as a partnership, where well-trained human medical personnel work with specialized robotic assistants to increase the quality and reliability of care.

Laser surgery on eyes is now close to complete machine control, and any activity where great precision is required is a candidate for machine operation. Machine diagnosis is trickier, and I suspect that skilled physicians will always be involved, but that they will be aided by dynamic, intelligent machines that can assess a large database of prior cases, medical records, medical knowledge, and pharmaceutical information. This assistance is already required, as the amount of information and the rapid addition of new information becomes overwhelming to practicing physicians. Moreover, as we get better diagnostic tools—more efficient analyses of body fluids and physiological records, DNA analyses, and various body scans—where some of the information will be routinely collected and sent from a patient's home or even place of work to the medical office, only a machine could keep up with the information. People are excellent at synthesis, at dynamic, creative decisions, at seeing the whole, global picture, whereas machines are superb at rapid search through large numbers of cases and information files, without being subject to the biases that accompany human memory. The team of trained physician and robotic assistant would be far superior to either working alone.

One common fear, of course, is that robots will take over many routine jobs from people, therefore leading to great unemployment and turmoil. Yes, more and more machines and robots will take over jobs, not only of lower-skilled workers, but increasingly of much routine work of all kinds, including some management. Throughout history, each new wave of technology has displaced workers, but the total result has been increased life span and quality of living for everyone, including, in the end, increased jobs—although of a different nature than before. In transitional periods, however, people are displaced and unemployed, for the new jobs that result often require skills very distant from those of the people who have been displaced. This is a major social problem that must be addressed.

In the past, most of the jobs replaced by automation have been low-level jobs, jobs that did not require much skill or education to perform. In the future, however, robots are apt to replace some highly skilled jobs. Will film actors be replaced by computer-generated characters that sound and act just as realistic, but are much more under the control of the director? Will robot athletes compete, if not with humans, then perhaps in their own leagues—but thereby leading to the demise of human leagues? Such a situation might very well happen with chess tournaments and leagues, now that computer chess players can beat even the best human players. What about jobs such as accounting, bookkeeping, drafting, stock keeping, or even simple management jobs? Will these be replaced? Yes, all this is possible; some of it has already started. Robot musicians? The list of potential activities is large, along with the dangers of social upheaval.

When robots are used for activities such as space exploration, dangerous coal mining, or search-and-rescue missions, or even when they do simple things around the house, such as vacuum cleaning and other chores, there is not apt to be much resistance. But when they start taking over large numbers of jobs or displacing large amounts of people from routine activities, then this does become a legitimate concern, one that raises serious issues for society.

I believe that we should welcome machines that eliminate the dreary tedium of many jobs—the dull shuffling of paperwork probably being even more demeaning than many of the low-paid, routine service jobs. This welcome, of course, assumes that machines will free people to engage in more creative activities, where they can apply their abilities both more pleasurably and effectively.

I have visited many parts of the world where poverty, continual hunger and starvation, and high death rates have made me doubt the benefits of today's systems. I have seen silk factories in India where young girls are locked into buildings, forced to weave from early morning till evening, locked in so that they cannot leave—or even escape the building if there is fire—without someone from the outside unlocking the doors. My study of history has taught me that such inequity, brutality, and callous treatment of so many is not unusual, and long predates the development of modern technology.

Yes, I see the downside of the deployment of intelligent machines and robots, but I also see the downside of no deployment. Call me an optimist, if you wish, but I believe that in the end, the human ingenuity that we show in creating these powerful devices will also serve us in ways to create more enriching, more enlightened activities for all of us. Optimism does not blind me to the inequities and problems of today's life: optimism reflects my belief that we can overcome them in the future. Yes, we still have poverty, starvation, political inequity, and wars, but these result more from the evils of people than from our technologies. I do not see why the introduction of smart, emotional robots and machines will change this situation, either for the worse or for the better. To change evil, we must confront it directly. It is a social, political, and human problem, not a technological one. This, of course, does not minimize this problem nor does it absolve us from working toward a solution. But the solution must be social and political, not technological.

The story becomes even more complex if I expand the view beyond the short-term horizon. At some point, robots and other machines are apt to become truly autonomous. This is a long time away, perhaps centuries, but it will happen. Then, there will indeed be major disruptions of life when much or all human work can be done by robots: farming, mining, manufacturing, distribution, and sales. Education and medicine. Even many aspects of art, music, literature, and entertainment. Robots may manufacture themselves. At that point, the relationship between natural animals and robots becomes exceedingly complex. The complexity will be amplified because many humans will actually be cyborgs—part human, part machine. Artificial implants already exist, mostly as medical prostheses; but some people are talking about having them implanted on demand, the better to enhance natural capabilities. Strength, athletic ability, sensory capability, memory, and decision making could all be aided by implanted, electronic, chemical, mechanical, biological, or nanotechnology devices. Steroids are used by athletes to enhance their existing strength, and laser treatment of the cornea has been done by some athletes and pilots to enhance normal acuity. The artificial lenses in my eyes—implanted after cataract removal—have provided me with far better vision than I have ever had before, with the sole problem being that my eyes cannot change their focus. But someday, artificial lenses will be able to focus, probably even better than natural ones, perhaps providing telescopic in addition to normal vision. When this happens, even people who do not have cataracts might wish to have their normal lenses replaced by these more effective ones. Even more striking artificial enhancement is possible. Such possibilities raise complex ethical issues, but these truly go beyond the boundaries of this book.

But this book does focus upon emotions and their role in the development of artificial devices and the way that human beings emotionally attach themselves to their belongings, their pets, and to one other. Robots might act like all of these. At first, robots will be belongings, but ones with clear personal attachment, for if a robot is with you for a large part of your life, able to interact, to remind you of your experiences, to give advice, or even just comic relief, there will be strong emotional attachments. Even today's robot pets, crude though they may be, have already evoked strong emotions among their owners. In the decades to come, robot pets may take on all the attributes of real pets and, in the minds of many people, be superior. Today people abuse and abandon their pets. Many communities have bands of stray cats or abandoned dogs scavenging. Might the same happen with robotic pets? Who is legally responsible for their care and maintenance? What if a robot pet injures someone? Who is legally responsible? The robot? The owners? The designer or manufacturer? With real pets, the owner is responsible.

And finally, what happens when robots act as independent, sentient beings, with their own hopes, dreams, and aspirations? Will something akin to Asimov's Laws of Robotics be necessary? Will they be sufficient? If robot pets can cause damage, what might an autonomous robot do; and if a robot causes damage, injury, or death, who is to blame, and what is the recourse? Asimov concluded in his novel *I, Robot*, that robots will indeed take over, that mankind will lose its own say in its future. Science fiction? Yes, but all future possibilities are fiction before they are fact.

We are in a new era. Machines are already smart, and they are getting smarter. They are developing motor skills, and soon they will have affect and emotion. The positive impact will be enormous. The negative consequences will also be significant. This is how it is with all technology: it is a two-edged sword always combining potential benefits with potential deficits.

EPILOGUE

We Are All Designers

I TRIED AN EXPERIMENT. I posted a request to some internet discussion groups for examples of products and web sites that they loved, hated, or had a love/hate relationship with. I received around 150 email responses, many passionate, and each listing several items. The responses were highly biased toward technology, not surprisingly, because this is the area in which most of the respondents worked; but technology did not receive high marks.

One of the problems with such a survey is the "too obvious to notice" effect, as reflected by the old folk tale that a fish is the last to notice water. Thus, if you ask people to describe what they see in the room in which they are sitting, they are apt to leave out the obvious: floor, walls, ceiling, and sometimes even windows and doors. People may not have reported what they truly liked because that might have been too close to them, too enmeshed in their lives. Similarly, they might have missed the disliked things because they were absent. Still, I found the responses interesting. Here are three examples:

Global chef's knives—beautiful, functional and simple. They are delightful to hold and use. I keep mine under my pillow (youch! just kidding).

The *pièce de résistance* is my watch. A George Jensen: sterling, large mirror face with two arms but no markings for numbers, the arm band is incomplete, only covers 3/4 of your wrist. Out of the ordinary, beautiful. (The design is in the Museum of Modern Art.) P.S. I stared at it, at least six years in Paris before I bought it.

My VW Bug: love it—it's simple, utilitarian, gets great gas mileage, small enough to park just about anywhere and just plain fun to drive. But I can't get past that stupid seat-lift-handle thing—it drives me bonkers. (The lift-the-seat handle on the front seats—they're in the "wrong" spot. Not one person has ever "gotten it right.")

Love it, hate it, indifferent to it. Our interaction with our everyday things reflects the three levels of design in very different ways. Loved objects ranged the gamut of all possible combinations of the three forms of design. Many an item was enjoyed solely for the visceral impact of its appearance:

After plunking down $400 for an iPod I almost wouldn't have cared about the product after having unwrapped the packaging, it was that nice. [The iPod is Apple Computer's music player.]

I bought a VW Passat because the controls inside the car were pleasurable to use and look at. (Get in one at night—the dashboard lights are blue and red-orange.) It makes driving more fun.

Remember the person in chapter 3 who bought water simply because the bottle looked so great? That response certainly belongs in this category:

I remember deciding to buy Apollinaris, a German mineral water, simply because I thought it would look so good on my shelves. As it turned out, it was a very good water. But I think I could have bought it even though it was not all that great.

Many products were loved for their behavioral design alone—that is, their function and utility, usability and understanding, and physical feel:

I like my OXO vegetable peeler, too. It handles eggplant, broccoli stems and anything else I throw at it. They make those nice comfy handles.

Lie-Nielsen hand planes: I can plane tiger maple and produce a smooth, glassy, surface where most planes would tear out chunks of wood.

Can opener: You may recall Victor Papanek's short book *How Things Don't Work*. In it he mentions a can opener. I finally found it a few years back—it's been reproduced by Kuhn Rikon as their LidLifter Can Opener. In brief, it opens the can by splitting the seam, rather than cutting through the top. Lots of reasons why that's a good thing, but it's an appliance I actually look forward to using. Hand operated, needs little cleaning, fits my hand, does its job, stores in a drawer, easily accessible. A dutiful servant, as a kitchen appliance should be.

The Screwpull lever model wine opener. Push down and pull up: the cork glides from the bottle. Push down again, squeeze and lift, and the cork comes off the corkscrew. It's wonderful! The day I got it I opened three bottles in a row, it was so much fun.

Reflective design also played a major role, with examples of trust, service, and just plain fun:

My Taylor 410 guitar. I trust my guitar. I know that it is not going to buzz when I play notes high on the fret board; it will stay in tune; the action on the neck will allow me to play chords and notes my hands cannot reach on other instruments.

I still tell people about my experience, years ago, with the Austin Four Seasons Hotel. I checked into my room to find a TV Guide on the bed, with a bookmark placed on the current date.

How about simply fun? I just got a souvenir mug; its decoration only becomes visible when it contains a hot drink, though: it's covered with heat-sensitive glazing that is dark purple-blue at room temperature and below but becomes transparent when hot. It's even practical: one look and I know when my coffee is no longer drinkable. Nice shape, too. I wanted it for the combination of all those factors, it's become my standard coffee mug now. Not exactly beautiful—but close.

Something that puts a smile on my face every time I visit the site is that the logo on the site "Google" is like a little cartoon that changes with relevance to something current. They will have a little devil peeking through the O for Halloween, or some snow caps on it during winter. I just love that.

Perhaps the most enthusiasm, though, was shown for communication services that enhanced social interaction and a sense of community. People loved their instant messenger tool:

I can't imagine my life without it.

Instant messenger is an integrated part of my life. With it I have a sense of connection to many of my friends and colleagues around the world. Without it, I feel as though a window to part of my world is bolted shut.

FIGURE EPI.1

The Google logo during the holiday season.

Google playfully transforms its logo during the end-of-the-year holiday season.

(Courtesy of Google.)

Email was seldom remarked upon—in part because it is like water to these technologists, but when it was spoken of, it was a love-hate response:

> I feel cut off from the civilized world if I don't get email. The volume of email I receive and feel obligated to answer almost puts this in the love/hate column, but on reflection, I may hate the volume, but I love the individual friends and family that comprise the unwieldy lot.

Household appliances and the personal computer (the pc) seem universally disliked: "Almost every appliance in my house is so ill-designed," said one. "Almost nothing about the PC is pleasurable," said another. And, remember, these respondents were technologists, most of them in the computer and internet business.

And, finally, some things were loved despite their faults. Hence, the love of the VW despite what the writer called "that stupid seat-lift-handle thing." Consider the following respondent's love for his espresso maker even though it was difficult to use (mind you, this response was from an expert on usability design). In fact, the lack of usability had some reflective appeal: "only a true expert, such as me, can use this properly."

> I love my espresso machine. Oddly not because of its ease-of-use (it hasn't got much!) but because it makes great coffee when you know how. It requires skill and the successful application of that skill is rewarding.

Over all, the responses showed that people can be passionate about their belongings, the services they use, and their experiences in life. Companies that provide extraordinary service reap the benefits: the special personal touch of being at a Four Seasons Hotel and finding on her bed the television guide, opened to the correct page, prompted that respondent to tell all her friends. Some people had bonded to their things: a guitar, their personal web site and the friends they had made through it, the feel of kitchen knives, a special rocking chair.

In my informal study I got at some aspects of our love and hate of things, but missed some of the truly loved items of the sort described by Csikszentmihalyi and Rochberg-Halton in their study *The Meaning of Things* that I discussed in chapter 2. They discovered such treasured items as a favorite set of chairs, family photographs, house plants, and books. Both of us ignored activities, such as our love or hate of cooking, sports, or class reunions. Both studies point to the development of true passion about particular items and activities in our lives—sometimes love, sometimes hate, but with strong emotional ties.

Personalization

How can mass-produced objects have personal meaning? Is it even possible? The attributes that make something personal are precisely the sorts of things that cannot be designed ahead of time, especially in mass production. Manufacturers try. Many provide customization services. Many allow special orders and specifications. And many provide a flexible product that, once it has been purchased, can be tuned and tailored by the people who use it.

Numerous manufacturers have tried to overcome the sameness of their product offerings by allowing customers to "customize" them. What this usually means is that the purchaser can choose the color or select from a list of accessories and extra-cost features. Cell phones can be equipped with different faceplates, so you can get one in different colors or designs—or paint it yourself. Some web sites advertise that you can design your own shoes, although, in fact, the only real alternatives you have are some choices among a fixed number of sizes, styles, colors, and materials (e.g., leather or cloth).

It is possible to have clothes made individually. In the past, they were made by tailors and seamstresses who would measure and fit a garment to your particular size and preferences. The result was well-fitting clothes, but the process is extremely slow, labor intensive, and, therefore, expensive. But what if technology were used to allow customization of everything—somewhat like the personal fit that one gets from tailors and seamstresses, but without the delay and cost? The idea is popular. Some believe that manufacturing to order—mass customization—will extend to everything: clothes, computers, automobiles, furniture. All would be manufactured specifically to specification: specify the configuration, wait a few days, and there it is. Several clothes manufacturers are already experimenting with the use of digital cameras to determine a person's measurements, lasers to cut the materials, and then computer-controlled manufacturing of the items. Some computer manufacturers already work this way, assembling products only after they have been ordered, allowing the customer to configure the product according to their desires. This has a benefit to the manufacturer as well: items are only manufactured after they have been purchased, which means that no stockpile of finished products is required, dramatically reducing the cost of inventory. When manufacturing processes are designed for mass-customization, individual orders can be made in hours or days. Of course, this form of customization is limited. You can't design a radically new form of furniture, automobile, or computer this way. All you can do is to select from a fixed set of options.

Are these customizations emotionally compelling? Not really. Yes, clothes might fit better, and the furniture might better suit some needs, but neither guarantees emotional attachment. Things do not become personal because we have selected some alternatives from a catalog of choices. To make something personal means expressing some sense of ownership, of pride. It means to have some individualistic touch.

We make our homes and places of work personal by the choice of items we place in them, how we arrange them, and how they are used. In the office, we arrange desk, table, and chairs, and post photographs, drawings, and cartoons on walls and doors.

Even items we dislike can provide a personal, redeeming touch: for example, a picture or a chair may be special because it is so detested—a legacy, perhaps, of a family member or a gift, and now there is no choice but to smile and keep it. Then, at family reunion after reunion, a family may fondly recall just how some disliked picture or chair once dominated a section of the house. It seems paradoxical, but the sharing of common negative feelings can lead to positive bonding of the participants: yesterday's hated object drives today's loved experience.

Determining a desirable arrangement of belongings is often more a process of evolution than of deliberate planning. We make continual small adjustments. We might move a chair a bit closer to the light and place the books and magazines we are reading near the chair. We bring over a table to hold them. Over time, the furniture and the belongings are adjusted to fit the inhabitants. The arrangement is unique to them and their activities. As activities and inhabitants change, so, too, does the arrangement of the house. Other people coming to live there may not necessarily find it fits their needs—it has become personal, it fits a person or a family—a quality that isn't transferable to others. Stuart Brand, in *How Buildings Learn*, has shown that even buildings change: as different occupants find their needs no longer met, they change the structure to meet their new needs, often changing an otherwise nameless, faceless building into a distinctive structure with personal value and meaning to its current inhabitants.

Objects themselves change. Pots and pans get banged and burned. Things are chipped and broken. But much as we may complain about marks, dents, and stains, they also make the objects personal—ours. Each item is special. Each mark, each burn, each dent, and each repair all contain a story, and it is stories that make things special.

While writing this book, I met with Paul Bradley, studio chief of IDEO, one of the largest industrial design firms in the United States. Bradley wanted to be able to design things that would reflect the experiences of an owner. He was searching for materials that would age gracefully, showing the dents and markings of use, but in a way that was pleasant and that would transform a store-bought, mass-produced item into a personal one, where the markings would add character and charm that was unique to the owner. He showed me a photograph of a pair of blue jeans, faded naturally through use, with a rectangular faded patch in the front pocket where the wearer had always kept his wallet. We discussed the bangs and markings on our own cooking utensils in our homes, and how they added to their appeal. We talked of favorite books made more comforting by the wear and marks of reading, enhanced through marginal notes and highlighting. And he showed me his Handspring Personal Digital Assistant (PDA)—which IDEO had designed—and told how he had deliberately dropped and banged it to see if the scuffs added personal history and charm (they didn't).

The trick is to make objects that degrade gracefully, growing old along with their owners in a personal and pleasurable manner. This kind of personalization carries huge emotional significance, enriching our lives. This is a far cry from the mass customization that allows a consumer to choose one of a fixed set of alternatives, but has little or no real personal relevance, little or no emotional value. Emotional value—now that is a worthy goal of design.

• • •

Customization

There is a tension between satisfying our needs by purchasing a ready-made object versus making it ourselves. Most of the time we are unable to build the objects we need, for we lack the tools and expertise, to say nothing of the time. But when we buy someone else's object, seldom does it fit our precise requirements. It is impossible to build a mass-produced item that fits every individual precisely.

There are five ways of dealing with this problem:

1. **Live with it.** Even if relatively inexpensive, mass-produced items are never quite what we need, we benefit from their lower cost.

2. **Customize.** Suppose everything was so flexibly designed that it could be modified as needed, wouldn't that solve the problem? The difficulty is that it is far more difficult to make something customizable than you might realize. Look at the modern computer software system, and you will immediately see the problem. My software offers a wide variety of customization options—so many that I can't even find them when I want them. So many that just learning how to customize is itself a daunting task. Moreover, these customizations invariably fail to satisfy. Everything I do is more complex because I must always choose among numerous alternatives. The things I really want to customize—my peculiar typing, spelling, and stylistic habits—can't be customized.

Proper customization does not come by further complicating an already complex system. No, proper customization comes about through combining multiple simple pieces. Invariably, if something is so complex that it requires the addition of multiple "preferences" or customization choices, it is probably too complex to use, too complex to be saved. I don't customize my pen; I do customize how I use it. I don't customize my furniture; I do customize through my choice of which piece to buy in the first place, where I put it, when I use it, and how.

3. Customized mass production. As I have just discussed, it is possible to have items manufactured to order. Customers get something configured to their tastes, and costs can be lower because there is no need for expensive supplies of unsold items. However, because the range of customization is limited to such things as choice of components, accessories, and color, this customization is far from personalization.

Still, this trend will continue. In the future, body parts, cases, and other parts of a design could be stamped, pressed, cut, or molded to order. Efficient assembly lines could put together customized structures. The choice of alternatives could expand. Manufacturing techniques are making it possible to extend the range of customization. This is the future.

4. Design our own products. In "the good old days," so it is said, we either made all our own things or went to the local craftsperson who would make something to our specifications, often as we watched. Some people still cherish those old days of folk arts—see, for example, John Seymour's wonderful description of them in his *Forgotten Arts and Crafts*. But as our needs get more complex and specialized in this ever-more technological, information-rich age, it is an impossible dream that many of us would possess the skills and time required to design and construct the objects required in everyday life. Nonetheless, it is not totally impossible to follow this route, and those who do reap many benefits. Some make their own clothing and construct furniture. Many people create and maintain gardens. Some even build their own airplanes or boats.

5. Modify purchased products. This is probably the favorite and most widely followed method to make purchased items into personal ones. Harley Davidson motorcycles are famous in this regard: people buy one from the factory and then immediately send it off to a custom detailer, who completely alters it, the alterations sometimes costing more than the cycle itself (already expensive). Each Harley is therefore unique, and owners pride themselves upon their special designs and paint jobs.

Similarly, building custom sound systems in automobiles is now a major business, with proud owners showing off their sound systems in regional meetings and contests. So, too, with customization of automobiles, changing the electronics that control the acceleration and performance, altering the shocks, the tires and rims, and paint.

Of course, the home is perhaps the biggest site of customization. Newly constructed, identical-looking houses soon transform themselves into individual homes as their occupants change furnishings, paint, window treatments, lawn, and, over years, modify the house's structure, adding rooms, changing garages, and so on.

We Are All Designers

> A space can only be made into a place by its occupants. The best that the designer can do is put the tools into their hands.
>
> —*Steve Harrison and Paul Dourish, "Re-place-ing space."*

We are all designers. We manipulate the environment, the better to serve our needs. We select what items to own, which to have around us. We build, buy, arrange, and restructure: all this is a form of design. When consciously, deliberately rearranging objects on our desks, the furniture in our living rooms, and the things we keep in our cars, we are designing. Through these personal acts of design, we transform the otherwise anonymous, commonplace things and spaces of everyday life into our own things and places. Through our designs, we transform houses into homes, spaces into places, things into belongings. While we may not have any control over the design of the many objects we purchase, we do control which we select and how, where, and when they are to be used.

Sit down and decide where to put your coffee cup, your pencil, the book you are reading, and the paper you wish to write on—you are designing. Even if this seems trivial and superficial, the essence of design is present: A set of choices, some better than others, perhaps none fully satisfactory. Possibly a dramatic restructuring to make everything work much better, but at some cost in effort, money, or even skills. Maybe if the furniture were rearranged or a new table purchased, the cup, pencil, book, and paper would fit much more naturally or the aesthetics would become more pleasurable? Once this is considered and a selection made, you are designing. Moreover, this activity is preceded by other designs; namely, the design of the building and the room, the selection of the furniture and its placement, and the location of the lights and their controls.

The best kind of design isn't necessarily an object, a space, or a structure: it's a process—dynamic and adaptable. Many a college student has made a desk by placing a flat-sided door on top of two filing cabinets. Boxes become chairs and book cases. Bricks and wood make shelves. Rugs become wall hangings. The best designs are the ones we create for ourselves. And this is the most appropriate kind of design—functional and aesthetic. It is design that's in harmony with our individual lifestyles.

Manufactured design, on the other hand, often misses the mark: Objects are configured and made according to particular specifications that many users find irrelevant. Ready-made, purchased items seldom fit our precise needs, although they might be close enough to be satisfactory. Fortunately, each of us is free to buy different items and then to combine them in whatever way works best for us. Our rooms fit our lifestyles. Our possessions reflect our personalities.

We are all designers—and have to be. Professional designers can make things that are attractive and that work well. They can create beautiful products that we fall in love with at first sight. They can create products that fulfill our needs, that are easy to understand, easy to use, and that work just the way we want them to. Pleasurable to behold, pleasurable to use. But they cannot make something personal, make something we bond to. Nobody can do that for us: we must do it for ourselves.

Personal web sites on the internet provide a powerful tool for people to express themselves, to interact with others all across the world, and to find communities that value their contributions. Internet technologies—such as newsletters, mailing lists, and chat rooms—allow people to congregate and share ideas, opinions, and experiences. Individual web sites and web logs allow personal expression, whether for art, music, photographs, or daily musing about events. These are all-powerful personal experiences that create strong emotional feelings. Here is how one person described her web site to me:

> My own web site—I sometimes want to give it up because it places great demands on my time, but it represents me online in such a personal way that it is impossible to imagine life without it. It brings me friends and adventures, travel and praise, humor and surprises. It has become my interface to the world. Without it an important part of me would not exist.

These personal web sites and web logs have become essential parts of many people's lives. They are personal, yet shared. They are loved and hated. They bring out strong emotions. These are truly extensions of the self.

Personal web sites, web logs, and other personal internet sites are prime examples of personal, nonprofessional design statements. Many people expend great amounts of time and energy in writing their thoughts, in collecting their favorite photographs, music, and video clips, and otherwise in presenting their personal face to the world. For many people, as with my correspondent, these personal statements represent them so intimately that it is inconceivable to imagine life without them—they have become an essential part of their self.

We are all designers—because we must be. We live our lives, encounter success and failure, joy and sadness. We structure our own worlds to support ourselves throughout life. Some occasions, people, places, and things come to have special meanings, special emotional feelings. These are our bonds, to ourselves, to our past, and to the future. When something gives pleasure, when it becomes a part of our lives, and when the way we interact with it helps define our place in society and in the world, then we have love. Design is part of this equation, but personal interaction is the key. Love comes by being earned, when an object's special characteristics makes it a daily part of our lives, when it deepens our satisfaction, whether because of its beauty, its behavior, or its reflective component.

The words of William Morris provide a fitting close to the book, just as they provided a fitting opening:

> If you want a golden rule that will fit everybody, this is it: Have nothing in your houses that you do not know to be useful, or believe to be beautiful.

Personal Reflections and Acknowledgments

IN SOME SENSE, this book is George Mandler's fault—surreptitiously sticking ideas in my head without my awareness. He hired me into the nascent Psychology Department of the University of California, San Diego, during the first year of the department's existence: the University itself had not yet graduated any students. Before I knew it, I had written a book (*Memory and Attention*) for his editorial series; developed an introductory textbook (*Human Information Processing*, written with Peter Lindsay), because of the course he had Peter and me teach; reconsidered my research on memory; and then entered the field of human error and accidents—whence my interest in design (under the philosophy that most human error is, in actuality, design error).

The Center for Human Information Processing—founded and run by Mandler—was host to the perceptual psychologist J. J. Gibson for a few summers, and these extended stays led to my many debates and continual disagreements with Gibson. These were delightful disagreements, enjoyed by both of us, disagreements of the most fruitful, scientific kind, the kind that teaches. The combination of my interest in errors and my adaptation of Gibson's notion of affordances led to *The Design of Everyday Things.* (Had Gibson not died, I'm certain he would still be arguing with me, disagreeing with my interpretation of his concept, ostentatiously disconnecting his hearing aid to show that he wasn't listening to my rebuttals, but secretly smiling and enjoying every minute.)

George was both a cognitive psychologist and a major figure in the study of emotion. But even though I spent many hours debating and discussing topics in emotion with him, reading all his works, I never knew quite how to integrate emotion into my studies of human cognition and, especially, into my studies of the design of products. I even gave a talk at the very first conference on cognitive science, in 1979, entitled "Twelve Issues for Cognitive Science," with emotion as number twelve. But even though I said we should study it, I didn't myself know how to go about doing it. My argument was convincing to at least one person in the audience: Andrew Ortony, now a professor at Northwestern University, tells me that he switched his area of research to emotion as a result of that talk.

In 1993, I left academia to join industry—serving as vice president at Apple Computer and then as an executive in other high-technology companies, including Hewlett Packard and an online, educational startup. In 1998, my colleague Jakob Nielsen and I established a consulting firm, the Nielsen Norman group, which has exposed me to a wide variety of products in several different industries. Eventually, though, academics drew me back, this time to the computer science department at Northwestern University. I now spend half time at the university, the other half with the Nielsen Norman group.

At Northwestern University, Andrew Ortony reawakened my dormant interest in emotion. In the decade that I was away from academia, much progress had been made in understanding the neuroscience and psychology of emotion. Moreover, while in industry, helping bring out a wide variety of products, from computers to appliances to web sites, I became sensitive to the powerful emotional impact that designs can produce. People were often far less interested in how well something worked, or even in what it did, than in how it looked and how it made them feel.

Together with William Revelle, a personality theorist in the department of psychology, Ortony and I decided to re-examine the literature on affect, behavior, and cognition to try to understand this emotional attraction. As our work progressed, it became clear that emotion and affect should not be separated from cognition—nor those from behavior, motivation, and personality; all are essential to effective functioning in the world. Our work forms the theoretical background for this book.

At roughly the same period, Bill Gross of Idealab! started a new company—Evolution Robotics—to create robots for the home. He asked me to join their advisory board; before long, I was deeply engrossed in the science of robots. Robots, I soon determined, need to have emotions to survive; indeed, emotions are essential for all autonomous creatures, humans or machine. To my pleasant surprise, I discovered that a research paper I had written in 1986 with the neuropsychologist Tim Shallice, on "will" as a control system, was being used in robotics. Aha! I began to see how it all fit together.

As these separate approaches coalesced, applications dropped out naturally. Our scientific explorations led us to propose that effective processing is best analyzed at three different levels. This insight clarified many issues. It soon became clear that many of the arguments about the role of emotion, beauty, and fun versus marketing concerns, advertising claims, and the positioning of products—along with the difficulties of making a product usable and functional—were often debates across the different levels of processing. All these issues are important, but all have different levels of influence, with different time courses, and at different places in the cycle of purchase and use.

My goal in writing this book is to put these apparently conflicting themes into one coherent framework based upon the three-level theory of affect, behavior, and cognition. With this framework I aim to provide a deeper understanding of the design process and the emotional impact of products.

So, thank you, George; thank you, Andrew; thank you, Bill.

THIS BOOK, like all my works, owes its existence to many other people. It started with the ever encouraging prods of both my patient agent, Sandy Dijkstra, and my business partner, Jakob Nielsen. No, not quite nagging, but continual reminders and encouragement. I'm always writing, always jotting things down, so out of these notes I created a manuscript entitled "The Future of Everyday Things." But when I tried teaching this material to students at Northwestern University, I discovered it lacked cohesion: the framework that tied the ideas together came from the new work on emotion that I was doing with Andrew Ortony and Bill Revelle, and this was not part of the book.

Ortony, Revelle, and I were developing a theory of emotion, and as we made progress, I realized that the approach could be applied to the field of design. Moreover, the work finally enabled me to resolve the apparent contradictions between my professional interest in making things usable and my personal appreciation of aesthetics. So I discarded that first book manuscript and started anew, this time using the theoretical work on emotion as a framework. Once again, I tried teaching the material, this time with far better success. My students in that first class, and then the ones who tried out the manuscript of this book, were all most helpful in transforming unrelated notes into coherent manuscript.

Along the way, my professional colleagues have provided considerable advice and resources. Danny Bobrow, my long-term colleague, with intelligent pokes, prods, and irritating questions where he would find the flaws in any argument I attempted. Jonathan Grudin, with a continuous flow of email, oftentimes from dawn to dusk, with comments, papers, and critiques. Patrick Whitney, head of the Institute of Design in Chicago, who invited me to serve on his board and provided both insightful comments and access to the industrial design community. Many of the faculty of the Institute of Design have been most helpful: Chris Conley, John Heskett, Mark Rettig, and Kei Sato. Nirmal Sethia, from California State Polytechnic University, Pomona, has been a continual source of contacts and information: Nirmal seems to know everyone in the field of industrial design and has made sure I was up-to-date.

THE POWERFUL team of interaction designers Shelley Evenson and John Rheinfrank always provide great insights (and John is a great chef). I thank Paul Bradley, David Kelly, and Craig Sampson of IDEO and Walter Herbst and John Hartman from Herbst LaZar Bell.

Cynthia Breazeal and Roz Picard from the MIT Media Laboratory provided numerous useful interactions, including visits to their laboratories, which contributed considerably to chapters 6 and 7. Rodney Brooks, head of the Artificial Intelligence Laboratory at MIT and a roboticist, was also a great source of information. Marvin Minsky, as always, provided much inspiration, especially with the manuscript of his forthcoming book, *The Emotion Machine.*

I tested many of my ideas on the several bulletin boards of the CHI community (the international society for Computer-Human Interaction), and many respondents have been most helpful. The list of correspondents is huge—hundreds—but I am especially indebted to my fruitful conversations with and suggestions from Joshua Barr, Gilbert Cockton, Marc Hassenzahl, Challis Hodge, William Hudson, Kristiina Karvonen, Jonas Löwgren, Hugh McLoone, George Olsen, Kees Overbeeke, Etienne Pelaprat, Gerard Torenvliet, and Christina Wodtke. I thank Kara Pernice Coyne, Susan Farrell, Shuli Gilutz, Luice Hwang, Jakob Nielsen, and Amy Stover of the Nielsen Norman group for their lively discussions.

Jim Stewart from Microsoft's XBOX division provided discussions of the game industry and the XBOX poster for my walls ("Go outside. Get some air. Watch a sunset. Boy, does that get old fast.").

The book slowly transformed from eighteen disorganized chapters into the present seven chapters, plus prologue and epilogue, through two massive rewrites, guided by Jo Ann Miller, my editor at Basic Books. She worked me hard—fortunately for you. Thanks, Jo Ann. And thanks to Randall Pink for diligently gathering final photographs and permissions.

Although I have left out many who helped during the long gestation period for this book, my thanks to all, named and unnamed, including all my students at both Northwestern University and the Institute of Design who helped me clarify my thoughts through the various revisions.

Don Norman
Northbrook, Illinois

Notes

Prologue: Three Teapots

3 "If you want a golden rule" (Morris, 1882. Quotation is from Chapter 3, "The Beauty of Life," originally delivered before the Birmingham Society of Arts and School of Design, February 19, 1880.)

6–7 "no new vehicle in recent memory has provoked more smiles" (Swan, 2002)

7 "It starts out with slight annoyance" (Hughes-Morgan, 2002)

12 "The neuroscientist Antonio Damasio studied people" (Damasio, 1994)

13 Parts of this chapter were published in *Interactions*, a publication of The Association for Computing Machines (Norman, 2002b)

Chapter One: Attractive Things Work Better

17 "two Japanese researchers, Masaaki Kurosu and Kaori Kashimura" (Kurosu & Kashimura, 1995)

17–18 "Japanese culture is known for its aesthetic tradition" (Tractinsky, 1997)

18 "So Tractinsky redid the experiment" (Tractinsky, 1997; Tractinsky, Katz, & Ikar, 2000)

18 "It requires a somewhat mystical theory" (Read, 1953, p. 61)

19 "The psychologist Alice Isen and her colleagues" (Ashby, Isen, & Turken, 1999; Isen, 1993)

21 “My studies of emotion, conducted with my colleagues” (Ortony, Norman, & Revelle, 2004)
31 “two words in the mythical language Elvish.” Tolkien’s books are, of course, well known (Tolkien, 1954a, b, c, 1956). This particular experiment was done in my classroom by Dan Halstead and Gitte Waldman (in 2002). They described the sound symbolism of Tolkien and, in a class demonstration, showed that people who had never heard Elvish could still reliably determine the meaning of its words.
32 “a sound symbolism governs the development” (Hinton, Nichols, & Ohala, 1994)

Chapter Two: The Multiple Faces of Emotion and Design

41 “almost 20,000 in the United States alone.” Magazine Publishers of America, figures for the year 2001.
http://www.magazine.org/consumer_marketing/index.html
46 “*kitsch.*” The Columbia Electronic Encyclopedia, Copyright © 1999, Columbia University Press. Licensed from Columbia University Press. All rights reserved. www.cc.columbia.edu/cu/cup/
46 “Nobody goes there anymore. It’s too crowded.” (Berra & Horton, 1989)
47 *The Meaning of Things* (Csikszentmihalyi & Rochberg-Halton, 1981)
48 “They are the first two chairs me and my husband bought.” (Csikszentmihalyi & Rochberg-Halton, 1981, p. 60)
48 “Household objects,” (Csikszentmihalyi & Rochberg-Halton, 1981, p. 187)
49 “when I walked through the exhibits on display at the San Francisco Airport.” The San Francisco Airport Museums: http://www.sfoarts.org/.
49 “The marvel of souvenir buildings” Text from the exhibit (Smookler, 2002)
52 “With every photo there is a story” (Cowen, 2002)
53 “Frohlich describes the possibilities this way” (Cowen, 2002)
54 “But put Asians in an individualistic situation” (Kitayama, 2002)
55 “the Heathkit Company.” The company has stopped making kits, although it still makes electronic learning materials. See the history at http://www.heathkit-museum.com/history.shtml.
55 “As the market researchers Bonnie Goebert and Herma Rosenthal put it” (Goebert & Rosenthal, 2001. The quotations are from Chapter 1: Listening 101, the value of focus groups.)
57 “Pirsig’s *Zen and the Art of Motorcycle Maintenance*” (Pirsig, 1974)
57 “The *American Heritage Dictionary* defines *fashion* . . .” *The American Heritage® Dictionary of the English Language, Fourth Edition,* Copyright © 2000 by Houghton Mifflin Company.
60 “Emotional branding is about building relationships.” Sergio Zyman, Former Chief Marketing Office of Coca-Cola. In the foreword to *Emotional Branding* (Gobé, 2001).
60 “Emotional branding is based on that unique trust that is established with an audience.” (Gobé, 2001. From the preface, page ix.)

Chapter Three: Three Levels of Design: Visceral, Behavioral, and Reflective

63 "I remember deciding to buy Apollinaris." Email from Hugues Belanger, in response to my query. May 6, 2002. To see the bottle, Belanger said: For a picture of the bottle, check out http://www.apollinaris.de/english/index.html (mouse over "Products" and click on "Apollinaris Classic").
64 "Walk down a grocery aisle." From the web site of "The Bottled Water Web": http://www.bottledwaterweb.com/indus.html.
64 "Package designers and brand managers." From the "Prepared Foods.com" web site: http://www.preparedfoods.com/archives/1998/9810/9810packaging.htm.
64 "Almost everyone who enjoys TyNant." From the TyNant web site: http://www.tynant.com/clients.htm.
69–70 "The principles of good behavioral design are well known . . . I laid them out in my earlier book, *The Design of Everyday Things*" (Norman, 2002a). Also see: (Cooper, 1999; Raskin, 2000).
74 "Understanding end-user unmet and unarticulated needs." From a Herbst LaZar Bell case study of the "Penguin" platform stepladder, sent to me mid-2002.
78 "Here, try this." The Tech Box is described in more detail in Tom Kelley's book about IDEO (Kelley & Littman, 2001, pages 142–146).
86 "Swatch Is Design" Student guide from the Swatch web site (Swatch Watch Corporation)
87 "In his important book about" (Coates, 2003, pp. 2.)
91 "The main goal in designing the Coaches Headset." Steve Remy, senior mechanical engineer and project manager for HLB, quoted in a press release for PTC's Pro/ENGINEER software that was used by HLB for the design. (July 23, 2001. Found at http://www.loispaul.com.)
92 "To the uninitiated" Copyright © 2002 by The New York Times Co. Reprinted with permission. (St. John, 2002)
93 "When you're wearing a thousand-dollar suit" (Rushkoff, 1999, p. 24)
95 "Paco Underhill's book" (Underhill, 1999)
96 "a wildly ambitious, hugely expensive science fiction allegory" and also, "symbolism ran such riot." From A. O. Scott's *New York Times* review of the restored movie (Scott, 2002).
98 "The brilliant conceptual artists Vitaly Komar and Alex Melamid" (Komar, Melamid, & Wypijewski, 1997)
98 "Perfectly 'user-centered design.'" Taken from Lieberman's essay "The Tyranny of Evaluation," available on his web site. I changed the phrase "user-centered interfaces" to "user-centered design" (with his permission) to make the point apply much more generally. (Lieberman, 2003)

Chapter Four: Fun and Games

99 "Professor Hiroshi Ishii of the MIT Media Laboratory runs back and forth."

Ishii's work is best seen at his web site: http://tangible.media.mit.edu/index.html. The bottles are described in (Ishii, Mazalek, & Lee, 2001; Mazalek, Wood, & Ishii, 2001).

99 "Imagine trying to play table tennis on a school of fish" (Ishii, Wisneski, Orbanes, Chun, & Paradiso, 1999)

100 "funology" (Blythe, Overbeeke, Monk, & Wright, 2003)

102 "the cook . . . would naturally be disappointed" (Ekuan, 1998, p. 18)

103 "A sense of beauty that lauds lightness" (Ekuan, 1998, pp. 79–81)

103 "with articles and books on 'positive psychology' and 'well-being' becoming popular" (Kahneman, Diener, & Schwarz, 1999; Seligman & Csikszentmihalyi, 2000; Snyder & Lopez, 2001)

103 "positive emotions *broaden* people's thought-action repertoires" (Fredrickson & Joiner, 2002). This quotation ends by referring the reader to other works by Fredrickson, in particular (Fredrickson, 1998, 2000).

104-105 "The book *Emotional Branding*" (Gobé, 2001)

105 "Patrick Jordan builds on the work of Lionel Tiger" (Jordan, 2000; Tiger, 1992)

108 "He picked up the hammer and ate it." See Coulson, King, & Kutas, 1998, although the particular example I made up isn't in this study.

109 "Pattern 134: Zen View" (Alexander, Ishikawa, & Silverstein, 1977, pp. 641–3)

109 "the parable of a Buddhist monk." I thank Mike Stone (at www.yawp.com) for reminding me of the parable. The quoted description of the parable comes from his discussion group posting.

111 "The seductive power of the design" (Khaslavsky & Shedroff, 1999, p. 45)

113 "On the napkin." Text accompanying the "collector's edition" of the juicer (Alessi, 2000).

113 "Fortunately, Khaslavsky and Shedroff have done the analysis for me." The lengthy quotation is from Khaslavsky & Shedroff, 1999, their Figure 1, p. 47 © 1999, Association for Computing Machinery, Inc. Reprinted by permission. (I have reformatted the text, but the words are the same.)

115 "Music plays a special role." An excellent review of these issues—and the source for this paragraph—is Krumhansl, 2002, especially note p. 46.

117 "All cultures have evolved musical scales." This section draws heavily from Krumhansl (2002) and Meyer (1956, p. 67).

120 "The designers of the Segway." Descriptive material for the "Segway Human Transporter." Amazon.com site. December 2002. Also, personal conversation with Dean Kamen, the inventor of Segway, February 25, 2003.

123 "the passions aroused in film" (Boorstin, 1990, p. 110)

124 "The vicarious eye puts our heart in the actor's body" (Boorstin, 1990, p. 110)

125 "the social scientist Mihaly Csikszentmihalyi" (Csikszentmihalyi, 1990)

126 "The word 'voyeur'" (Boorstin, 1990, p. 12)

126 "The voyeur's eye demands." Boorstin (1990), pages 13, 61, and 67.

127 "It can ruin the most dramatic moment" (Boorstin, 1990, p. 13)

129–130 "Overslept, woke at 8:00." From an interview with Will Wright, game developer of the Sims, conducted by Amazon.com Computer Game Editor Mike Fehlauer: http://www.playcenter.com/PC_Games/interviews/will_wright_the_sims.htm.
131 "Video games were once thought" (Klinkenborg, 2002)
131 "In *The Medium of the Video Game*" (Wolf, 2001). The list of categories was taken from the excerpt at: http://www.robinlionheart.com/gamedev/genres.xhtml.
132 "As Verlyn Klinkenborg says" (Klinkenborg, 2002)

Chapter Five: People, Places, and Things

136 "Byron Reeves and Clifford Nass" (Reeves & Nass, 1996)
136 "B. J. Fogg shows how people think." The table is taken from his Table 5.1. (Fogg, 2002)
138 "It starts out with slight annoyance" (Hughes-Morgan, 2002)
139–140 "Now we get into the complex emotions" The basic analysis presented here comes from the work of the psychologists Andrew Ortony, Gerald Clore, and Allan Collins (Ortony, Clore, & Collins, 1988), although I have modified their interpretation somewhat, to fit the special emphasis on design in this book. The modifications are also in line with the work that I have done with them, in particular with Andrew Ortony and William Revelle. (Ortony, Norman, & Revelle, 2004)
141 "My 10-inch Wusthof chef knife." Email received in response to my query on the CHI discussion group. May 2002. (CHI is the Computer-Human Interaction society.)
143 "It's human nature to trust our fellow man" (Mitnick & Simon, 2002, p. 32.)
144 "social psychologists Bibb Latané and John Darley" and "Bystander apathy" (Latané & Darley, 1970)
145 "Crew Resource Management" (Wiener, Kanki, & Helmreich, 1993)
146 "As I was writing this book" (Hennessy, Patterson, Lin, & National Research Council Committee on the Role of Information Technology in Responding to Terrorism, 2003)
148 "Everywhere is nowhere." Thanks to John King, Dean of the School of Information at University of Michigan for the quotation from Seneca.
150 "Instant messenger." Responses to my request to an on-line discussion group on design to tell me products they love or hate (Dec. 2002). The two paragraphs in this example were written by different people.
151 "Vernor Vinge, one of my favorite" (Vinge, 1993)
154 "Attention span . . . ten seconds." I believe it is in James's *Principles of Psychology* (James, 1890), but although I have relied on this quotation for more than thirty years, it is also more than thirty years since I read it. Try as I might, I have been unable to find it again in order to provide a proper bibliographic reference.
154 "We carve out our own private spaces." See William Whyte's book *City: Rediscovering the Center* (Whyte, 1988).
157 "Continually divided attention." Linda Stone, then a vice president of Microsoft. Personal communication, PopTech! Conference, Camden, ME, 2002.

Chapter Six: Emotional Machines

161 "Dave, stop . . ." Excerpt from the movie *2001*, from Bizony (1994), p. 60.

163 Photograph of C3PO and R2D2. *Star Wars: Episode IV—A New Hope* © 1977 and 1997 Lucasfilm Ltd. & ™. All rights reserved. Used under authorization. Unauthorized duplication is a violation of applicable law.

165 "The psychologists Robert Sekuler and Randolph Blake" (Sekuler & Blake, 1998)

166 "as happens to some emotionally impaired people" (Damasio, 1994, 1999)

169 "The 1980s was the decade of the PC." Toshitada Doi, president of Sony Digital Creatures Laboratory. (Nov. 2000)

170–171 "Neal Stephenson's science fiction novel" (Stephenson, 1995)

173 "Rodney Brooks, one of the world's leading roboticists" (Brooks, 2002). The quotation is from page 125.

175 "Masahiro Mori, a Japanese roboticist": *The Buddha in the Robot* (Mori, 1982). The argument that we are more bothered when the robot is too close to human appearance comes from an essay by Dave Bryant (Bryant, not dated). Bryant attributes the argument to Mori, but when I bought Mori's book and read it, although I enjoyed the book, I found not even a hint of this argument. Nonetheless, it is a great point.

176 "Philip K. Dick's" (Dick, 1968)

180 "I realized it would be a heck of a lot easier if we just gave them emotions." Picard's quote comes from Cavelos (1999, pp. 107–108), but she reaffirmed it to me during my visit to her laboratory in 2002.

183 "The extent to which emotional upsets can interfere with mental life" (Goleman, 1995). The quote was taken from (Kort, Reilly, & Picard, 2001).

184 "Professor Rosalind Picard" (Picard, 1997)

185 "Even the most controlled person." The basic research was done by Paul Ekman (Ekman, 1982, 2003). An excellent popular description is in the New Yorker article by Malcolm Gladwell (Gladwell, 2002).

186 "National Research Council" (National Research Council Committee to Review the Scientific Evidence on the Polygraph, 2002)

188 "Perhaps the earliest such experience was with Eliza." The work on Eliza was done in the 1960s. It is reviewed in Weizenbaum's book (Weizenbaum, 1976).

190 "Do you think that I can use" Conversation between Danny Bobrow, Eliza, and the VP. A transcript of the conversation has been made available by Güven Güzeldere and Stefano Franchi: I copied it from their website (Güzeldere & Franchi, 1995). I also confirmed the details through conversation and email with Bobrow (Dec. 27, 2002).

191 "*Computer Power and Human Reason*" (Weizenbaum, 1976)

191 "*Designing Sociable Robots*" (Breazeal, 2002)

192 (Figure 6.6). The photograph of Kismet comes from http://www.ai.mit.edu/projects/sociable/ongoing-research.html (with permission). For more detailed description, see Cynthia Breazeal's book *Designing Sociable*

Robots (Breazeal, 2002).
194 "These things push on our buttons." Quotations of Turkle taken from an interview with L. Kahney, in Wired.com (but I corrected the grammar). (Kahney, 2001).

Chapter Seven: The Future of Robots

196 "most of the world governments banned robot use" (Asimov, 1950)
197 "Asimov's Four Laws of Robotics." Roger Clarke, in his writings and on his authoritative web site (Clarke, 1993, 1994), dates the origins of laws one, two, and three from Clarke's discussion with science fiction author and editor John Campbell in 1940 (Asimov, 1950, 1983). The zeroth law was added 45 years later, in 1985 (Asimov, 1985).
200 "Do what I mean, not what I say." Note that DWIM (Do What I Mean) is a very old concept: Warren Teitelman introduced it into the command interpretation system of the LISP computer programming system in 1972. When it works, it is very, very nice.
202 "Asimov's main failure" A good review of the work on emergent systems, that is, against central control, is in Johnson's book *Emergence* (Johnson, 2001).
203 "There already are some safety regulations that apply to robots" (*Industrial Robots and Robot System Safety. Occupational Safety and Health Administration, US Department of Labor, OSHA Technical Manual (TED 1-0.15A)*, 1999)

Epilogue: We Are All Designers

215 "You may recall Victor Papanek's short book" (Papanek & Hennessey, 1977)
220 "Stuart Brand . . . has shown" (Brand, 1994)
223 "John Seymour's wonderful description" (Seymour, 2001)
224 "Steve Harrison and Paul Dourish" (Harrison & Dourish, 1996)
226 "My own web site." Response to my query for people on an email discussion list abut design to tell me of products or websites they love, hate, or have a love-hate relationship with (Dec. 2002).
227 "If you want a golden rule" (Morris, 1882. Quotation is from chapter 3, "The Beauty of Life," originally delivered before the Birmingham Society of Arts and School of Design, February 19, 1880.)

References

Alessi, A. (2000). Creating Juicy salif. Product brochure accompanying the Special Anniversary Edition 2000 of the Juicy Salif. Crusinallo, Italy: Alessi.

Alexander, C., Ishikawa, S., & Silverstein, M. (1977). *A pattern language: Towns, buildings, construction*. New York: Oxford University Press.

Ashby, F. G., Isen, A. M., & Turken, A. U. (1999). A neuropsychological theory of positive affect and its influence on cognition. *Psychological Review, 106*, 529–550.

Asimov, I. (1950). *I, Robot*. London: D. Dobson. (Reprinted numerous times; see: Asimov, I. [1983]).

Asimov, I. (1983). *The Foundation trilogy: Foundation, Foundation and empire, Second foundation; The stars, like dust; The naked sun; I, robot*. New York: Octopus/ Heinemann.

Asimov, I. (1985). *Robots and empire* (1st ed.). Garden City, NY: Doubleday.

Berra, Y., & Horton, T. (1989). *Yogi: It ain't over*. New York: McGraw-Hill.

Bizony, P. (1994). *2001: Filming the Future*. London: Arum Press.

Blythe, M. A., Overbeeke, K., Monk, A. F., & Wright, P. C. (2003). *Funology: From usability to enjoyment*. Boston: Kluwer Academic Publishers.

Boorstin, J. (1990). *The Hollywood eye: What makes movies work*. New York: Cornelia & Michael Bessie Books.

Brand, S. (1994). *How buildings learn: What happens after they're built*. New York: Viking.

Breazeal, C. (2002). *Designing sociable robots*. Cambridge, MA: MIT Press.

Brooks, R. A. (2002). *Flesh and machines: How robots will change us*. New York: Pantheon Books.

Bryant, D. (not dated). The uncanny valley: Why are monster-movie zombies so horrifying and talking animals so fascinating? Retrieved, 2003, http://www.arclight.net/~pdb/glimpses/valley.html.

Cavelos, J. (1999). *The science of Star Wars* (1st ed.). New York: St. Martin's Press.

Clarke, R. (1993). Asimov's laws of robotics: Implications for information technology, Part 1. IEEE Computer, 26 (12), 53–61. http://www.anu.edu.au/people/Roger.Clarke/SOS/Asimov.html.

Clarke, R. (1994). Asimov's laws of robotics: Implications for information technology, Part 2. IEEE Computer, 27 (1), 57–66. http://www.anu.edu.au/people/Roger.Clarke/SOS/Asimov.html.

Coates, D. (2003). *Watches tell more than time: Product design, information, and the quest for elegance*. New York: McGraw-Hill.

Cooper, A. (1999). *The inmates are running the asylum: Why high-tech products drive us crazy and how to restore the sanity*. Indianapolis: Sams; Prentice Hall.

Coulson, S., King, J. W., & Kutas, M. (1998). Expect the unexpected: Event-related brain response to morphosyntactic violations. *Language and Cognitive Processes*, 13 (1), 21–58.

Cowen, A. (2002, June). Talking photos: Interview with David Frohlich. *mpulse, a Cooltown magazine*. http://www.cooltown.com/mpulse/0602-thinker.asp.

Csikszentmihalyi, M. (1990). *Flow: The psychology of optimal experience*. New York: Harper & Row.

Csikszentmihalyi, M., & Rochberg-Halton, E. (1981). *The meaning of things: Domestic symbols and the self*. Cambridge, UK: Cambridge University Press.

Damasio, A. R. (1994). *Descartes' error: Emotion, reason, and the human brain*. New York: G. P. Putnam.

Damasio, A. R. (1999). *The feeling of what happens: Body and emotion in the making of consciousness*. New York: Harcourt Brace.

Dick, P. K. (1968). *Do androids dream of electric sheep?* (1st ed.). Garden City, NY: Doubleday.

Ekman, P. (1982). *Emotion in the human face* (2nd ed.). Cambridge, UK: Cambridge University Press.

Ekman, P. (2003). *Emotions revealed: Recognizing faces and feelings to improve communication and emotional life*. New York: Henry Holt & Co./Times Books.

Ekuan, K. (1998). *The aesthetics of the Japanese lunchbox*. Cambridge, MA: MIT Press.

Fogg, B. J. (2002). *Persuasive technology: Using computers to change what we think and do*. New York: Morgan Kaufman Publishers.

Fredrickson, B. L. (1998). What good are positive emotions? *Review of General Psychology, 29*, 300–319.

Fredrickson, B. L. (2000). Cultivating positive emotions to optimize health and well-being. Prevention & Treatment (an electronic journal), 3 (Article 0001a). Available on-line with commentaries and a response at http://journals.apa.org/prevention/volume3/toc-mar07–00.html

Fredrickson, B. L., & Joiner, T. (2002). Positive emotions trigger upward spirals toward emotional well-being. *Psychological Science*, *13* (2), 172–175.

Gladwell, M. (2002, August 5). Annals of Psychology: The naked face: Can experts really read your thoughts? *The New Yorker*, 38–49.

Gobé, M. (2001). *Emotional branding: The new paradigm for connecting brands to people*. New York: Allworth Press.

Goebert, B., & Rosenthal, H. M. (2001). *Beyond listening: Learning the secret language of focus groups*. New York: J. Wiley. URL for Chapter 1: Listening 101: The value of focus groups. http://www.wileyeurope.com/cda/cover/0,,0471395625%7Cexcerpt,00.pdf.

Goleman, D. (1995). *Emotional intelligence*. New York: Bantam Books.

Güzeldere, G., & Franchi, S. (1995). Constructions of the mind: Dialogues with colorful personalities of early AI. Stanford Electronic Humanities Review, 4 (2). http://www.stanford.edu/group/SHR/4–2/text/dialogues.html.

Harrison, S., & Dourish, P. (1996). *Re-place-ing space: The role of place and space in collaborative systems*. ACM. Proceedings of the Conference on Computer Support of Collaborative Work (CSCW). New York: ACM.

Hennessy, J. L., Patterson, D. A., Lin, H. A., & National Research Council Committee on the Role of Information Technology in Responding to Terrorism (Eds.). (2003). *Information technology for counterterrorism: Immediate actions and future possibilities*. Washington, DC: The National Academies Press.

Hinton, L., Nichols, J., & Ohala, J. J. (1994). *Sound symbolism*. Cambridge, UK: Cambridge University Press.

Hughes-Morgan, M. (2002, February 25). Net effect of computer rage. This is London, http://www.thisislondon.com/dynamic/news/story.html?in_review_id=506466&in_review_text_id=469291.

Industrial Robots and Robot System Safety. Occupational Safety and Health Administration, US Department of Labor, OSHA Technical Manual (TED 1–0.15A). (1999). http://www.osha.gov/SLTC/machineguarding/publications.html. Section V entitled "Control and Safeguarding Personnel" outlines specific means for safeguarding robot systems.

Isen, A. M. (1993). Positive affect and decision making. In M. Lewis & J. M. Haviland (Eds.), *Handbook of emotions* (pp. 261–277). New York: Guilford.

Ishii, H., Mazalek, A., & Lee, J. (2001). Bottles as a minimal interface to access digital information. Computer Human Interaction (CHI–2001), Extended Abstracts. ACM Press http://tangible.media.mit.edu/papers/Bottles_CHI01/Bottles_CHI01.pdf.

Ishii, H., Wisneski, C., Orbanes, J., Chun, B., & Paradiso, J. (1999). PingPongPlus: Design of an athletic-tangible interface for computer-supported cooperative play. Pittsburgh, PA. CHI 99: Conference on Human Factors in Computing Systems.

http://tangible.media.mit.edu/papers/PingPongPlus_CHI99/PingPongPlus_CHI99.html

James, W. (1890). *Principles of psychology.* New York: Holt.

Johnson, S. (2001). *Emergence: The connected lives of ants, brains, cities, and software.* New York: Scribner.

Jordan, P. W. (2000). *Designing pleasurable products: An introduction to the new human factors.* London: Taylor & Francis.

Kahneman, D., Diener, E., & Schwarz, N. (1999). *Well-being: The foundations of hedonic psychology.* New York: Russell Sage Foundation.

Kahney, L. (2001). Puppy love for a robot. Wired news. http://www.wired.com/news/culture/0,1284,41680,00.html.

Kelley, T., & Littman, J. (2001). *The art of innovation: Lessons in creativity from IDEO, America's leading design firm.* New York: Currency/Doubleday.

Khaslavsky, J., & Shedroff, N. (1999). Understanding the seductive experience. Communications of the ACM, 42 (5), 45–49. http://hci.stanford.edu/captology/Key_Concepts/Papers/CACMseduction.pdf.

Kitayama, S. (2002). *Cultural psychology of the self: A renewed look at independence and interdependence.* Stockholm, 2000. Proceedings of the XXVII international congress of psychology. Vol. II. Psychology Press. http://www.hi.h.kyoto-u.ac.jp/users/cpl/thesis/k2.pdf.

Klinkenborg, V. (2002, December 16). Editorial observer; Living under the virtual volcano of video games this holiday season. *The New York Times,* Section A, pp. 26.

Komar, V., Melamid, A., & Wypijewski, J. (1997). *Painting by numbers: Komar and Melamid's scientific guide to art.* New York: Farrar Straus Giroux.

Kort, B., Reilly, R., & Picard, R. W. (2001). An affective model of interplay between emotions and learning: Reengineering educational pedagogy—building a learning companion. ICALT–2001 (International Conference on Advanced Learning Technologies).

Krumhansl, C. L. (2002). Music: A link between cognition and emotion. *Current Directions in Psychological Science, 11* (2), 45–50.

Kurosu, M., & Kashimura, K. (1995, May 7–11). Apparent usability vs. inherent usability: experimental analysis on the determinants of the apparent usability. Denver, Colorado. *Conference companion on human factors in computing systems.* 292–293.

Latané, B., & Darley, J. M. (1970). *The unresponsive bystander: Why doesn't he help?* Englewood Cliffs, NJ: Prentice-Hall.

Lieberman, H. (2003). The Tyranny of Evaluation. Retrieved, 2003, http://web.media.mit.edu/~lieber/Misc/Tyranny-Evaluation.html.

Mazalek, A., Wood, A., & Ishii, H. (2001, August 12–17). *GenieBottles: An interactive narrative in bottles.* Proceedings of SIGGRAPH. ACM Press. http://tangible.media.mit.edu/papers/genieBottles_SG01/genieBottles_SG01.pdf.

Meyer, L. B. (1956). *Emotion and meaning in music.* Chicago: University of Chicago Press.

Mitnick, K. D., & Simon, W. L. (2002). *The art of deception: Controlling the human element of security.* Indianapolis: Wiley.

Mori, M. (1982). *The Buddha in the robot* (S. T. Charles, Trans.). Boston: Charles E Tuttle Co.

Morris, W. (1882). Hopes and fears for art: Five lectures delivered in Birmingham, London, and Nottingham, 1878–1881. London: Ellis & White. http://etext.library.adelaide.edu.au/m/m87hf/chap3.html. (Quotation is from chapter 3, "The Beauty of Life," originally delivered before the Birmingham Society of Arts and School of Design, February 19, 1880.)

National Research Council Committee to Review the Scientific Evidence on the Polygraph. (2002). *The polygraph and lie detection.* Washington, DC: National Academies Press.

Norman, D. A. (2002a). *The design of everyday things.* New York: Basic Books. (The reissue, with a new preface, of *The psychology of everyday things.*)

Norman, D. A. (2002b). Emotion and design: Attractive things work better. *Interactions Magazine*, ix (4), 36–42. http://www.jnd.org/dn.mss/Emotion-and-design.html

Ortony, A., Clore, G. L., & Collins, A. (1988). *The cognitive structure of emotions.* Cambridge, UK: Cambridge University Press.

Ortony, A., Norman, D. A., & Revelle, W. (2004). The role of affect and proto-affect in effective functioning. In J.-M. Fellous & M. A. Arbib (Eds.), *Who needs emotions? The brain meets the machine.* New York: Oxford University Press.

Papanek, V. J., & Hennessey, J. (1977). *How things don't work* (1st ed.). New York: Pantheon Books.

Picard, R. W. (1997). *Affective computing.* Cambridge, MA: MIT Press.

Pirsig, R. M. (1974). *Zen and the art of motorcycle maintenance.* New York: Bantam Books.

Raskin, J. (2000). *The humane interface: New directions for designing interactive systems.* Reading, MA: Addison Wesley.

Read, H. E. (1953). *Art and industry, the principles of industrial design* (3rd. ed.). London: Faber and Faber.

Reeves, B., & Nass, C. I. (1996). *The media equation: How people treat computers, television, and new media like real people and places.* Stanford, CA: CSLI Publications (and New York: Cambridge University Press).

Rushkoff, D. (1999). *Coercion: Why we listen to what "they" say.* New York: Riverhead.

Scott, A. O. (2002, July 12). Critic's notebook: A restored German classic of futuristic angst. *The New York Times,* B, pp. B18. http://www.nytimes.com/2002/07/12/movies/12METR.html.

Sekuler, R., & Blake, R. (1998). *Star Trek on the brain: Alien minds, human minds.* New York: W. H. Freeman. http://www2.shore.net/~sek/STontheBrain.html.

Seligman, M. E. P., & Csikszentmihalyi, M. (2000). Positive psychology: An introduction. *American Psychologist, 55* (1), 5–14.

Seymour, J. (2001). *The forgotten arts & crafts.* New York: Dorling Kindersley.

Smookler, K. (2002). Text from the San Francisco Airport Museums exhibit on Miniature Monuments: email.

Snyder, C. R., & Lopez, S. J. (Eds.). (2001). *Handbook of positive psychology.* New York: Oxford University Press.

St. John, W. (2002, July 14). A store lures guys who are graduating from chinos. *The New York Times,* Sunday Styles, pp. 9–1, 9–8. http://www.nytimes.com/2002/07/14/fashion/14JEAN.html.

Stephenson, N. (1995). *The diamond age, or, A young lady's illustrated primer.* New York: Bantam Books.

Swan, T. (2002, Sunday, June 2). Behind the wheel/Mini Cooper: Animated short, dubbed in German. *The New York Times,* Automobiles, pp. 12.

Swatch Watch Corporation. Swatch basics: Facts & figures from the world of Swatch [Internet (PDF) White paper]. Retrieved, December 2002, http://www.swatch.com/fs_index.php?haupt=collections&unter=.

Tiger, L. (1992). *The pursuit of pleasure.* Boston: Little Brown.

Tolkien, J. R. R. (1954a). *The fellowship of the ring: being the first part of The lord of the rings* (Vol. pt. 1). London: George Allen & Unwin.

Tolkien, J. R. R. (1954b). *The lord of the rings.* London: Allen & Unwin.

Tolkien, J. R. R. (1954c). *The two towers: being the second part of The lord of the rings* (Vol. pt. 2). London: G. Allen & Unwin.

Tolkien, J. R. R. (1956). *The return of the king: being the third part of The lord of the rings* (Vol. v. 3). Boston: Houghton Mifflin.

Tractinsky, N. (1997). Aesthetics and apparent usability: Empirically assessing cultural and methodological issues. *CHI 97 Electronic publications: Papers* http://www.acm.org/sigchi/chi97/proceedings/paper/nt.htm.

Tractinsky, N., Katz, A. S., & Ikar, D. (2000). What is beautiful is usable. *Interacting with Computers, 13* (2), 127–145.

Underhill, P. (1999). *Why we buy: The science of shopping.* New York: Simon & Schuster.

Vinge, V. (1993). *A fire upon the deep.* New York: Tor.

Weizenbaum, J. (1976). *Computer power and human reason: From judgment to calculation.* San Francisco: W. H. Freeman.

Whyte, W. H. (1988). *City: Rediscovering the center* (1st ed.). New York: Doubleday.

Wiener, E. L., Kanki, B. G., & Helmreich, R. L. (1993). *Cockpit resource management.* San Diego: Academic Press.

Wolf, M. J. P. (2001). The medium of the video game (1st ed.). Austin: University of Texas Press. See "Genre and the Video Game" at http://www.robinlionheart.com/gamedev/genres.xhtml.

Index